How is the staggering biodiversity of the parasitoid insects maintained? This book explores patterns in host–parasitoid interactions, including parasitoid community richness, the importance of parasitoids as mortality factors and their impact on host densities as determined by the outcomes of parasitoid introductions for biological control.

It documents general patterns using data sets generated from the global literature and evaluates potential underlying biological, ecological and evolutionary mechanisms. A theme running throughout the book is the importance of host refuges as a major constraint on host–parasitoid interactions. Much can be learnt from the analysis of broad patterns; a few simple rules can go a long way in explaining the major components of these interactions.

This book will be an invaluable resource for researchers interested in community ecology, population biology, entomology and biological control.

PATTERN AND PROCESS IN HOST–PARASITOID INTERACTIONS

PATTERN AND PROCESS IN HOST–PARASITOID INTERACTIONS

BRADFORD A. HAWKINS

NERC Centre for Population Biology, Imperial College,
Silwood Park, Ascot, UK

CAMBRIDGE UNIVERSITY PRESS
Cambridge, New York, Melbourne, Madrid, Cape Town, Singapore, São Paulo

Cambridge University Press
The Edinburgh Building, Cambridge CB2 2RU, UK

Published in the United States of America by Cambridge University Press, New York

www.cambridge.org
Information on this title: www.cambridge.org/9780521460293

First published 1994
This digitally printed first paperback version 2005

A catalogue record for this publication is available from the British Library

Library of Congress Cataloguing in Publication data

Hawkins, Bradford A.
Pattern and process in host–parasitoid interactions / Bradford A.
Hawkins.
p. cm.
Includes bibliographical references (p.) and index.
ISBN 0 521 46029 8 (hc)
1. Parasitic insects. 2. Insects – Parasites. 3. Parasitoids.
4. Host–parasite relationships. 5. Insect–plant relationships.
I. Title
QL496.H3 1994
595.7′05249 – dc20 93-45350 CIP

ISBN-13 978-0-521-46029-3 hardback
ISBN-10 0-521-46029-8 hardback

ISBN-13 978-0-521-01944-6 paperback
ISBN-10 0-521-01944-3 paperback

Contents

Preface

In January 1986 I arrived at the University of York in northern England as an NSF-sponsored post-doctoral fellow to conduct a field experiment involving insects on bracken. One day in February, while waiting to begin the real work in the summer, John Lawton stopped me in the hallway and asked me if I thought it would be possible to use the parasitoid–host literature to examine patterns in the number of parasitoid species that individual herbivore species support, in much the same way as he and others had analyzed herbivore species richness patterns on plants. My response was immediate and self-assured: No! Everyone knows that the records are too fragmented and the data too piecemeal and biased to have any chance of finding any meaningful results. But after returning to my office and thinking it over for a while, I decided that it could not hurt to try, particularly since the Yorkshire winter is ideal for this type of research. The following day I made the first of what was to become many trips to the library to search the literature.

Looking back, I suspect that my response to John's question may have been overly pessimistic. The initial analysis of the parasitoids of British insects revealed a number of potentially interesting patterns, but it raised far more questions than it answered. A follow-up analysis of North American patterns of parasitoid diversity produced results surprisingly similar to those found in Britain, but, again, there were nagging doubts about what they really meant. In a continuing effort to test and expand the results, the research snowballed. Now, seven years later, with my reputation as a data miner firmly established, and my papers on this topic beginning to clutter-up the literature, the time seems appropriate to publish in a single work the sum of the interrelated projects that have used this comparative approach and to synthesize the results. This is what this monograph does.

As with any such project of this scale, it is important to acknowledge my friends and colleagues who have contributed data and ideas that have shaped the final outcome. First, I thank those with whom I have collaborated at various stages in the research: Richard Askew, Raymond Gagné, Paul Gross, Michael Hochberg, John Lawton and Mark Shaw. I have learned a great deal from them. Correspondence, discussion and occasionally heated debate with a number of other workers have also been stimulating and invaluable; particularly helpful have been Ian Gauld, Charles Godfray, Nick Mills, John Noyes, Peter Price, William Sheehan, Chris Thomas and Jeff Waage. I also thank my wife, Ada, who spent untold hours in front of photocopy machines generating grist for the analytical mill. I am especially grateful to the staff of the National Lending Library, Boston Spa (UK) for their patience, and to the CAB International Institute of Biological Control for providing a copy of their biological control data base. Phases of this research were supported financially by the National Science Foundation, the Leverhulme Foundation and the NERC Centre for Population Biology. Finally, I thank Paul Gross, John Lawton, Gabor Lövei, Nick Mills, Peter Price, Mark Shaw, William Sheehan and Teja Tscharntke for their trenchant critiques of all or parts of this manuscript. All errors of fact or interpretation that remain are entirely their fault!

Bradford A. Hawkins

1
Introduction

Insect parasitoids, their phytophagous hosts and their hosts' foodplants compose a major proportion of the world's biodiversity. Based on the current numbers, organisms directly involved in plant–insect–parasitoid interactions represent well over half of all known species (Price, 1980; Strong, Lawton & Southwood, 1984). Of course, the bulk of species remains undescribed, but even if and when a complete global inventory of species is available, plant–insect diversity will still dominate the biota (World Conservation Monitoring Centre, 1992). Given the staggering number of species involved, host–parasitoid interactions can be expected to be exceedingly complex, and virtually any conceivable type of relationship between parasitoids and their hosts can probably be found to exist. The massive literature that has already been generated provides just a taste of complexities remaining to be discovered, and dealing with this diversity presents population and community ecologists with a daunting task. If we accept that understanding host–parasitoid interactions is an important undertaking worth the time and resources it will clearly require, what are our options for achieving this understanding?

One extreme is to pursue a case-specific approach, in which a few model systems are developed, and the fine details worked out through observation and experimentation. If we pick our systems carefully, this will provide us with a strong foundation for understanding at least some of the range of possible interactions. On the other hand, there will always be questions about the generality of the results, and intense arguments among workers about the nature of host–parasitoid interactions will undoubtedly emerge, with positions usually reflecting the particular system on which favored theories are based. Perhaps more importantly, it will leave the vast proportion of systems essentially unstudied, since there will never be enough researchers, resources or time to study intensively more than an infinitesimal fraction of the communities that exist. Despite these drawbacks, if we persist long enough with this approach we may ultimately accumulate enough data to start generating reasonably synthetic theories about the major forces that drive host–parasitoid interactions, in maybe 100 years' time.

An alternative approach is to step back from detailed studies of model systems and search for pattern in data sets composed of large numbers of cases, in the hope that general relationships do exist and can be identified. It does not logically follow from this approach that the details are unimportant, only that much can be learned from broadly based analyses of pattern. Once these patterns have been documented, we can then proceed with more detailed studies to evaluate specific mechanisms and test the practical value of the generalizations. The research reported herein represents this latter approach in its extreme form.

The study of host–parasitoid interactions began in earnest in the late 1880s, following the successful introduction of the vedalia beetle for the biological control of the cottony cushion scale in California. Subsequently, intense interest in the use of natural enemies to control insect pests led to a plethora of studies of insects and their parasitoids, with the emphasis very much on the potential of parasitoids as regulatory agents. Beginning in the 1960s, the emphasis began to shift toward studying the parasitoid complexes of non-pest species as well as those of pests, partially in the hope that the lessons learned would improve biological control and pest management and partially because of increasing interest in achieving a basic understanding of how communities are structured. Since then, the parasitoids of a relatively large number of host species have been studied in both agricultural and natural habitats.

As a consequence of 100+ years of research, an extensive primary literature on parasitoids and their hosts has been generated. Despite this wealth of information, no one has ever attempted to compile the existing data to determine what, if any, conclusions can be reached about the nature of the interactions. Of course, we still have data on a very small proportion of host–parasitoid systems (see Fig. 2.1, p. 11 for a graphic illustration of just how pitifully studied some parts of the world are), and further, many of the studies have been rather cursory. Accepting that the data are grossly incomplete and of highly variable quality (and always will be), do we currently have enough information at hand to identify general patterns and evaluate the processes that underpin them?

I have three main goals in this monograph. First, I simply document a large number of patterns in host–parasitoid interactions, constrained by what I am able to address with the data at my disposal. Over the past seven years I have published a number of such analyses, each based on parts of an ever growing collection of data. This has resulted in a somewhat fragmented presentation of many of the patterns. An important function of this work is to pull together the many phases of the research and to present the full scope of the results in a consistent context. For this reason, the text is dominated by the presentation of the patterns, supported by statistical evaluations of their significance. Documenting the full range of patterns that actually exists represents the essential first step for building a synthetic theory of host–parasitoid interactions that has a chance of reasonably approximating reality.

As implied by the title, I also go beyond the patterns and attempt to evaluate the mechanisms that might account for them. Trying to infer process from pattern is extremely risky (Keddy, 1989), but to restrict the scope of this monograph to presenting page after page of patterns without trying to understand why they exist would be intellectually sterile, not to mention exceedingly boring to read. As is the case with all pattern analyses, it will not always be possible to identify unambiguously the forces that give rise to particular patterns, and in many cases manipulative experiments or additional data will be required before we can even sort out the range of possibilities. But I have not let this deter me from trying to identify mechanisms, since it is only by generating one set of hypotheses that alternatives can then be formulated and the appropriate experiments designed and/or the observational data collected to test them. Despite some claims to the contrary (Peters, 1991), identifying the mechanisms underlying patterns in nature is the only way forward for the development of ecology as a science. Also, I do not want to get involved in the philosophical debate about the relative merits of different approaches in ecology. Attempts to explain the results of pattern analyses are sometimes said to represent 'weak inference' at best (e.g. Hairston, 1989), and it is true that interpretations must be cautious. But the technique I have employed whenever possible relies on generating predictions from hypotheses and then examining the data to see if the predicted patterns exist. After reading this monograph, the reader is free to judge the extent that this approach has been successful and that it is possible to identify at least the major forces that underpin host–parasitoid interactions using a comparative methodology.

My second goal is to expand and refine previously established patterns and to present new ones. Despite the number of papers that I have already published using this approach, many results remain unpublished or have been substantially improved and expanded since their initial publication. This monograph contains relatively few results extracted directly from my papers; instead, it focuses on new, more complete analyses and considers several issues that have not been previously raised.

Third, and most importantly, it has become apparent during this work that the three main aspects of host–parasitoid interactions that I am able to quantify, namely the species richness of parasitoid complexes, parasitoids as mortality agents, and the outcomes of classical biological control introductions, all represent facets of a single problem, and there may be a few, relatively simple rules that can conceptually unify parasitoid community ecology, insect population biology and the applied practice of biological control. In addition to mechanisms evaluated to explain specific patterns, a theme running throughout this monograph is the importance of refuges from parasitism as a general explanation for a wide range of patterns, a mechanism I refer to here as the 'susceptibility hypothesis'. The single most important purpose of this work is to develop a simple, yet apparently robust, hypothesis for a number of important ecological problems, both basic and applied.

The monograph is divided into chapters, but its basic structure is similar to that of a research paper, and there is a great deal of interdependence among the chapters. In Chapter 2, I present how the data were generated and analyzed. Chapters 3–6 discuss different aspects of host–parasitoid interactions, beginning with the analysis of total parasitoid community richness (Chapter 3) and then dividing the parasitoid complexes into biological and taxonomic components to permit more detailed analyses and better evaluations of potential underlying mechanisms (Chapter 4). Chapter 5 shifts the emphasis from parasitoid diversity to patterns in host mortality and biological control, and Chapter 6 extends the analyses to obligatory hyperparasitoid complexes. In the final chapter, I bring together the major patterns in order to construct a simple conceptual model for the structure and dynamics of host–parasitoid interactions using host feeding biology as the template. I then evaluate several alternative hypotheses that have been proposed to account for at least some of the patterns and discuss theoretical and empirical support for the hypothesis that links host feeding biology to refuges from parasitism. Finally, I present an analysis of the outcomes of specific parasitoid introductions for biological control to demonstrate how the susceptibility hypothesis may be useful for making predictions in particular host–parasitoid systems.

2
Data and methodology

2.1 Introduction

All of the results reported in this monograph are based on analyses of data sets generated by extracting information from an extensive global literature. Because the bulk of the data was originally collected and reported by workers for reasons unrelated to how I use them, a proper evaluation of the patterns and their significance hinges on an understanding of the structure of the data sets, their scope, and their limitations. This chapter provides details on data collection and analytical methodology. I first describe the data sets on which the analyzes are based, representing primary parasitoid and hyperparasitoid assemblage sizes, levels of parasitoid-induced host mortality, and the outcomes of biological control introductions using parasitoids. I then define the independent and dependent variables actually analyzed and, finally, I briefly describe the statistics used.

2.2 Parasitoid species richness

The data quantifying parasitoid assemblage size, defined as the number of parasitoid species per herbivorous host species, consist of lists of parasitoid species recorded from individual host species from four orders of holometabolous herbivorous insects (Coleoptera, Diptera, Lepidoptera and Hymenoptera). Herbivory is defined in its broad sense and includes insects that feed on either dead or living plant tissues, as well as those feeding on fungus-infested tissues. Generally, lists were generated from primary literature sources, with a few exceptions. Only hosts studied within their presumed native ranges were included. The basic 'sampling' scheme comprised visits to various libraries (University of York, University of Leeds, Imperial College, The Natural History Museum (British Museum), and The National Lending Library (Boston Spa), all in the UK, and the University of Texas and Texas

A & M University in the USA). Any journals held by each library which were believed likely to contain relevant papers were searched completely from approximately 1930, or Volume 1, until the most recent available issue. Additional published and unpublished data were obtained by direct correspondence with researchers known to be active in the study of parasitoids.

Two separate species richness data sets were generated. The first includes primary and facultatively hyperparasitic parasitoids attacking host larvae and pupae. Parasitoids of eggs or adults were excluded, largely because they have been rather sporadically studied. Undoubtedly, a few egg-larval parasitoids have been inadvertently included when they belong to higher taxa that do not appear to be universally egg-larvals and when workers did not report that parasitoids reared from larvae actually attack eggs. These represent only a tiny fraction of the parasitoid records.

The second data set represents obligatory hyperparasitoids and includes both true hyperparasitoids that attack the primary parasitoid through the herbivore, and pseudohyperparasitoids that attack the primary parasitoid in the cocoon stage after it has emerged from the host (Shaw & Askew, 1976b).

Separate parasitoid lists for each herbivore species were generated by country for smaller countries and by state or province for larger countries (i.e. the USA, Canada, Brazil, India and China). For the former USSR, separate lists were generated for each republic, except for Russia, where lists were generated by administrative region (Oblast). When a host species had been studied in several smaller countries, the country with the longest parasitoid list was used; when a host had been studied in several regions within a large country, the region with the most parasitoids was used. This methodology has two effects. First, it reduces the influence of differences in intensity of study, since the vast majority of the hosts included in the data set have been studied once in a particular country or in the individual states/provinces/republics of North America, China, India or the USSR. Second, it reduces species-area effects by decreasing the area over which a host has been studied in large countries to within the range of areas in which hosts have been studied in smaller countries. Given the ubiquity of species-area effects in nature, there is little doubt that they are important to parasitoid diversity as well (Hawkins & Lawton, 1987). However, for the bulk of parasitoid complexes in my data set, there is not enough information to incorporate area into the analyses. Therefore, it was decided at an early stage to attempt to control for area in so far as possible. An analysis incorporating the geographic distributions of hosts would be interesting and is needed, but it is beyond the scope of my global parasitoid diversity data set. The goal of this study was not to list all of the parasitoids reported from every herbivore species whenever and wherever studied. Instead, the lists represent a sample of the parasitoid complexes taken from the enormous literature on host–parasitoid associations, with an effort to make them as comparable as possible.

An additional source of variation which was more difficult to deal with arises from potential temporal variation in parasitoid species richness. Some studies were conducted for only one year or host generation, whereas others lasted several years or generations. Because of extreme variability in the way that parasitoid lists are reported in the literature, it was not possible to quantify temporal variation in assemblage size in most cases, so I did not try. It is assumed that this has increased the variability within the data but has not introduced any systematic errors. This represents a second aspect of parasitoid species richness that deserves closer inspection.

Finally, herbivores with no reported parasitoids were excluded, because it was frequently difficult to judge whether an absence of parasitoids was real, or just that workers did not make concerted efforts to rear them. I estimate that fewer than 50 cases of valid 0s were excluded under this criterion, so this decision has had a minimal influence on the results.

Under the above constraints, the data set comprises the parasitoid complexes attacking 1289 host species in at least 108 families from all of the world's major biogeographical regions (Table 2.1), and hosts from over 85 (pre-1990) countries are represented (Figs. 2.1, 2.2). The data set is not an exhaustive compilation of all of the parasitoid complexes that have been documented, particularly because many extensive secondary literature sources were excluded. But it does include a wide range of host and parasitoid taxa sampled over most of the world.

2.3 Host mortality

Parasitoid-induced host mortality was estimated by extracting percentage parasitism data from the sources used to generate the parasitoid diversity data. Total, combined percentage apparent parasitism was recorded whenever workers provided information. When studies encompassed several host populations or were done over several host generations, the maximum mortality level for any single population/generation was used. Similarly, when multiple studies were available for an individual host species, the study reporting the highest percentage parasitism was used (which in a few cases was not the same study used to generate the species richness data). Finally, whenever possible only percentage parasitism of host larval stages was used, because the foundation of the analyses rests on the feeding habits of host larvae. However, for hosts that support mostly or exclusively larval-pupal parasitoids (e.g. many tephritids), or which pupate in the same location where larvae feed, mortality estimates from pupal collections were included.

Table 2.1. *Taxonomic distribution of host species by biogeographic region.*

Order	Family	NEAR	NEOT	PALE	AFRO	ORIE	AUST	Total
Coleoptera	Anthribidae	–	–	–	3	–	–	3
	Apionidae	6	1	16	16	1	–	40
	Bostrichidae	1	1	3	–	–	–	5
	Bruchidae	8	1	8	28	2	–	47
	Buprestidae	5	3	5	1	2	1	17
	Cerambycidae	27	2	14	8	6	3	60
	Chrysomelidae	17	5	18	2	8	5	55
	Coccinellidae	–	1	4	–	1	–	6
	Curculionidae	35	12	52	3	19	5	126
	Elateridae	2	–	–	–	–	–	2
	Languriidae	1	–	–	–	–	–	1
	Mordellidae	–	–	–	–	1	–	1
	Nitidulidae	–	–	3	–	–	–	3
	Platypodidae	–	–	–	3	–	–	3
	Rhynchitidae	1	–	–	–	–	–	1
	Scarabaeidae	1		7	4	4	2	18
	Scolytidae	19	1	67	1	–	–	88
	Tenebrionidae	–	–	4	–	–	–	4
Diptera	Agromyzidae	18	8	28	5	10	–	69
	Anthomyiidae	5	1	20	–	3	–	29
	Cecidomyiidae	77	13	88	5	34	–	217
	Chloropidae	2	–	10	1	–	–	13
	Diopsidae	–	–	–	6	–	–	6
	Ephydridae	1	–	1	–	1	–	3
	Fergusoninidae	–	–	–	–	1	–	1
	Lonchaeidae	–	2	1	–	–	–	3
	Micropezidae	–	–	–	–	1	–	1
	Opomyzidae	–	–	1	–	–	–	1
	Otitidae	–	1	–	–	–	–	1
	Psilidae	–	–	1	–	–	–	1
	Stratiomyiidae	–	–	–	–	–	1	1
	Syrphidae	–	–	1	–	–	–	1
	Tephritidae	36	12	23	8	12	10	101
	Tipulidae	–	–	2	–	–	–	2
Lepidoptera	Acrolepidae	–	–	2	–	–	–	2
	Adelphocephalidae	–	1	–	–	–	–	1
	Agaristidae	–	–	–	–	–	1	1
	Agonoxenidae	–	–	–	–	–	2	2
	Anthelidae	–	–	–	–	–	2	2
	Apatelodidae	–	1	–	–	–	–	1
	Arctiidae	5	3	3	2	9	–	22
	Blastoblasidae	–	–	–	–	1	–	1
	Bombycidae	–	–	–	–	2	–	2
	Castniidae	–	1	–	–	–	–	1
	Choreutidae	–	–	1	–	–	–	1
	Cochylidae	1	–	2	–	–	–	3

Table 2.1. (*cont.*)

Order	Family	NEAR	NEOT	PALE	AFRO	ORIE	AUST	Total
Lepidoptera *cont.*								
	Coleophoridae	2	1	7	–	2	–	12
	Cosmopterigidae	3	–	3	–	–	–	6
	Cossidae	–	–	3	–	1	–	4
	Cryptophasidae	–	–	–	–	1	–	1
	Dalcoridae	–	1	–	–	–	–	1
	Eriocranidae	–	–	3	–	–	–	3
	Eupterotidae	–	–	–	–	–	1	1
	Gelechiidae	25	2	8	2	12	–	49
	Geometridae	19	3	16	2	8	4	52
	Glyphipterigidae	–	–	–	–	1	–	1
	Gracillariidae	14	2	37	1	7	–	61
	Heliodinidae	–	1	–	–	–	–	1
	Heliozelidae	1	–	1	–	–	–	2
	Hepialidae	2	–	–	–	–	2	4
	Hesperiidae	–	1	2	1	2	–	6
	Hyblaeidae	–	–	–	–	1	–	1
	Incurvariidae	6	–	–	–	–	1	7
	Lasiocampidae	6	–	10	6	2	–	24
	Limacodidae	–	6	1	2	16	1	26
	Lycaenidae	1	1	6	–	3	–	11
	Lymantriidae	4	–	12	6	7	1	30
	Lyonetiidae	4	–	8	3	2	–	17
	Megalopygidae	2	–	–	–	–	–	2
	Momphidae	1	–	–	–	–	–	1
	Nepticulidae	4	–	10	–	–	–	14
	Noctuidae	36	9	30	18	32	3	128
	Notodontidae	5	–	2	5	–	–	12
	Nymphalidae	–	13	7	4	–	1	25
	Oecophoridae	6	2	2	1	1	3	15
	Papilionidae	2	1	3	–	2	2	10
	Pieridae	2	1	6	–	2	–	11
	Plutellidae	–	–	3	–	–	–	3
	Psychidae	1	5	2	3	8	3	22
	Pterophoridae	8	1	2	–	4	–	15
	Pyralidae	64	24	18	10	56	3	175
	Riodinidae	–	1	–	–	–	–	1
	Saturniidae	7	4	–	6	–	–	17
	Sesiidae	2	–	5	–	1	–	8
	Sphingidae	2	2	–	–	2	–	6
	Thaumetopoeidae	–	–	4	–	–	–	4
	Thyrididae	–	–	–	1	1	–	2
	Tischeriidae	1	–	1	–	–	–	2
	Tortricidae	70	3	86	8	11	2	180
	Yponomeutidae	6	–	18	–	3	–	27
	Zygaenidae	1	–	2	–	1	–	4

Table 2.1. (*cont.*)

Order	Family	NEAR	NEOT	PALE	AFRO	ORIE	AUST	Total
Hymenoptera	Argidae	2	–	3	–	3	1	9
	Cephidae	3	–	5	–	–	–	8
	Cynipidae	12	–	43	–	–	–	55
	Diprionidae	18	1	6	–	5	–	30
	Eulophidae	–	–	–	1	–	–	1
	Eurytomidae	4	1	9	4	1	1	20
	Pamphiliidae	1	–	3	–	–	–	4
	Pergidae	–	–	–	–	–	2	2
	Perilampidae	–	–	–	–	–	2	2
	Pteromalidae	2	–	–	–	–	1	3
	Siricidae	5	–	5	–	4	–	14
	Syntexidae	1	–	–	–	–	–	1
	Tanaostigmatidae	–	–	–	–	1	–	1
	Tenthredinidae	13	1	39	–	3	–	56
	Torymidae	–	–	2	–	–	–	2
	Xiphydriidae	1	–	2	–	–	–	3
	Xyelidae	2	–	–	–	–	–	2
Unidentified species		–	–	2	–	1	–	3
Total		639	157	822	180	325	66	2189

NEAR, Nearctic; NEOT, Neotropics; PALE, Palearctic; AFRO, Afrotropics; ORIE, Oriental; AUST, Australasia.

It should be realized that the measurement of host mortality due to natural enemies is subject to a wide range of sampling problems (Simmonds, 1949; Petersen, 1986), and very few of the papers consulted to generate the mortality data used the detailed techniques that have been proposed to overcome these problems (Van Driesche, 1983; Van Driesche *et al.*, 1991). Thus, the mortality data are subject to extensive, but unquantifiable, measurement error. The data also include measures of mortality based solely on parasitoid emergence from field samples. These estimates may exclude additional sources of parasitoid-induced mortality that arise from host feeding (DeBach, 1943; Jervis & Kidd, 1986), or when the host is killed but parasitoids fail to emerge (e.g. Petersen, Catangui & Watson, 1991; Sandanayake & Edirisinghe, 1992). The absolute levels of mortality are often very crude estimates, but if they contain no strong systematic biases they provide a useful, albeit noisy picture of patterns of mortality and are suitable for comparisons with other data sets.

It is also worth noting why I use maximum parasitism rates rather than some measure of central tendency. The rationale of generating this data set was to determine how much mortality parasitoids are capable of inflicting on their

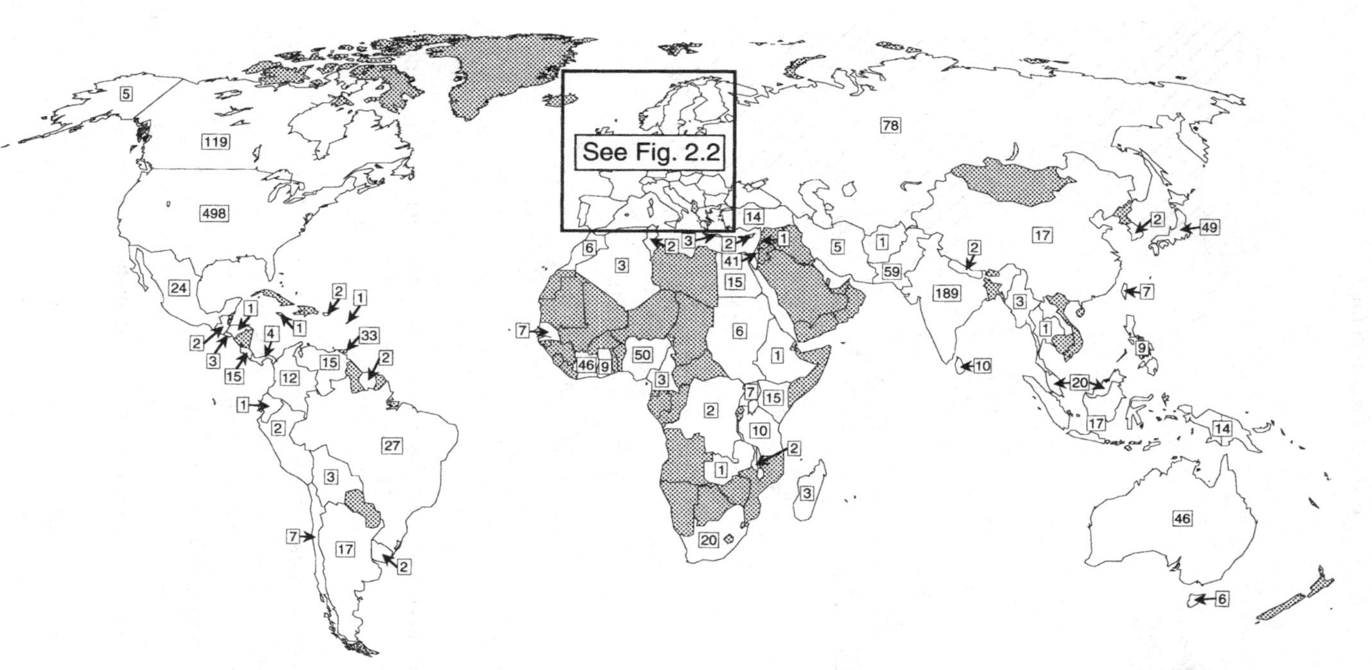

Fig. 2.1. Geographic distribution of the host–parasitoid complexes comprising the parasitoid species richness data set. Numbers are the numbers of host species in each country/region. Areas contributing no host species to the data are shaded.

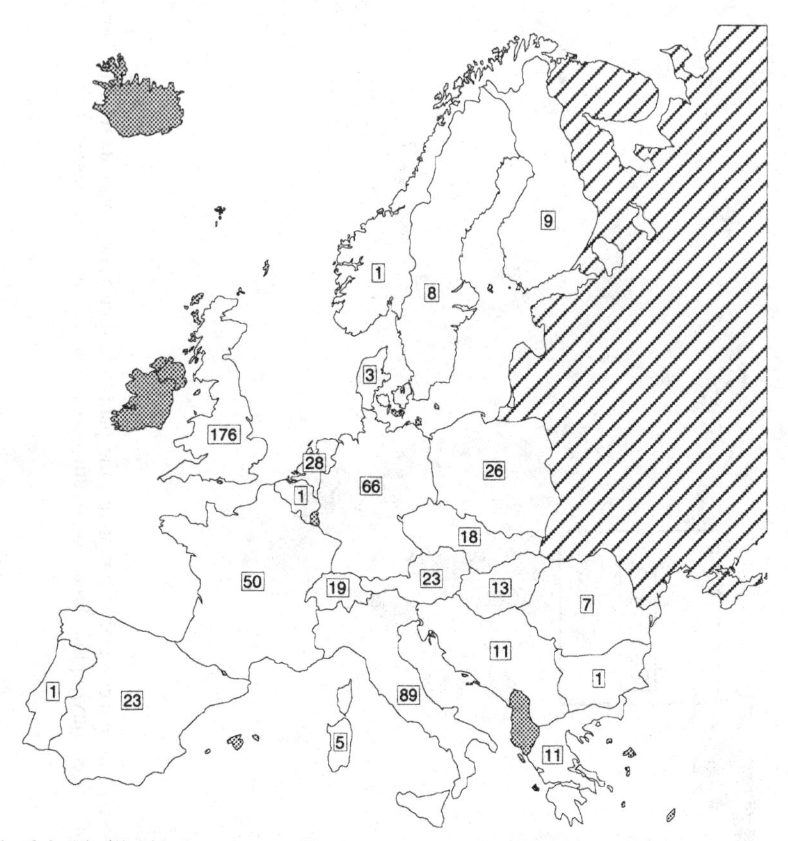

Fig. 2.2. Distribution by country of host species contributing parasitoid species richness data for western and central Europe. Numbers are the number of host species. Areas contributing no host species are shaded.

hosts. Parasitism rates can vary for many reasons, only a few of which reflect the innate ability of parasitoids to attack their hosts. If, for example, one host population is found to suffer 90% parasitism, whereas another suffers only 10%, what is the susceptibility of that host species to attack? To claim that it is 50% (the mean) is misleading, since we have direct evidence that parasitoids are capable of parasitizing at least 90% of their hosts when conditions are favorable (for whatever reason). The higher figure is certainly an overestimate of the realized levels of mortality in all host populations all of the time, but it does provide the best estimate of what parasitoids are capable of. It should be realized that this is consistent with the way the species richness data were compiled. Within the defined geographic limits outlined above, when given a choice for a host species studied in several regions with parasitoid lists in each region having different lengths, the longest list was used. Thus, both data sets measure maximum values, not averages.

2.4 Parasitoid impact

There is no doubt that parasitoids can have significant impacts on host populations, although it has proven difficult to generalize about when parasitoids may or may not be important (Price, 1987). To examine potential patterns of the impact of parasitoids on host densities, I use the outcomes of biological control introductions against insect pests. The data were derived from the BIOCAT biological control data base compiled by the CAB International Institute of Biological Control (Greathead, 1986; Waage, 1990; Greathead & Greathead, 1992), which contains over 4000 records of natural enemy introductions throughout the world. For each introduction of a larval or pupal parasitoid for the control of a pest belonging to one of the four orders encompassed by my study, biological and taxonomic information concerning the host and the outcome of the introduction (unknown, parasitoid failed to establish, parasitoid established but no control observed, or some degree of control achieved) was recorded. These outcomes are then used to estimate success rates for different types of parasitoid–host systems.

As biological control workers have repeatedly pointed out (Keller, 1984; Waage, 1990; Greathead & Greathead, 1992) catalogs of biological control projects are highly imperfect records of the outcomes of parasitoid introductions on host populations. Probably the most severe potential problem is evaluating how meaningful a recorded 'success' really is. My approach has been simply to treat all levels of success, whether partial, substantial or complete, as equal evidence that parasitoids have had an observable impact on host densities, without attempting to quantify how severe the impact might have been.

2.5 Independent variables

The number of potential biological, ecological and evolutionary factors that could drive host–parasitoid interactions is large. Unfortunately, most of these potential influences have not been measured for more than a few systems. Therefore, only a very limited subset can be included in an analysis based on a collection of field studies. I have included seven general types of independent variables, the selection of which depended on a combination of *a priori* expectations that they might be important and on the ability to extract them from a large proportion of the studies. These variables are:

(1) Host feeding niche Eight types of hosts are distinguished, based on how and where they feed as larvae: (a) external folivores, those herbivores which feed fully exposed on the leaf surface; (b) leaf rollers/webbers/tiers/tent builders, including all exophytic feeders which construct a simple shelter; (c) casebearers; (d) leaf/needle miners; (e) gallers; (f) borers of stems, bark, flowers, pods, or fruits; (g) root feeders, including both endophytic and exophytic feeders which share the characteristic of feeding underground; and (h) mixed exophytic/endophytic feeders, defined as those herbivores which switch between exophytic and endophytic feeding during larval development or which are exposed when moving from one endophytic feeding site to another. Typically, herbivores were assigned to feeding classifications based on how they feed after the first instar of larval development. For example, many Lepidoptera that eventually roll leaves mine them as first instars, but all such herbivores were classified as leaf rollers. Similarly, the mixed category includes only those herbivores which switch feeding modes or become partially mobile from the second instar onward. A potential problem with any attempt to classify into discrete groups the feeding habits of organisms as diverse as insect herbivores is that the groupings must be somewhat subjective. Undoubtedly, others would classify some of the herbivores differently from my groups, particularly when I have had to make a decision from minimal biological information. I believe that although some 'errors' are present, they represent a very small proportion of the data.

(2) Host taxon Hosts were classified by order and family. Host taxonomy is a reflection of evolutionary relationships and serves as a proxy variable for evaluating the importance of phylogenetic inertia. Detailed study of the strict taxonomic relationships among parasitoids and their hosts is beyond the scope of this work. These variables are used primarily to identify specific exceptions to general patterns and to determine the extent that taxonomic differences confound interpretation of the results.

(3) Host generic diversity Similar to host taxon, this variable measures the potential importance of evolutionary factors (in this case host diversification patterns on a shorter time scale than host order or family). Because the taxonomy of most herbivorous insect groups is grossly incomplete and in need of revision, it was possible to generate this variable only for Great Britain. For these hosts, the number of species recorded in Britain in each herbivore genus was recorded. The classifications of Kloet & Hincks (1972, 1976, 1977, 1978) were used.

(4) Sample size The numbers of hosts and parasitoids reared were recorded when workers provided explicit information. This variable represents primarily an 'artefact' which is used to ensure that sampling biases do not account for the major patterns. However, sample sizes might also be expected to be loosely correlated with host abundance, and it is probable that patterns based on sample size may include non-artefactual, biologically meaningful components.

(5) Climate Two temperature-related variables are used as indicators of the general climatic conditions with which hosts and their parasitoids must contend. The main function of these variables is to permit tropical and extra-tropical comparisons and to provide more direct measures of local climatic conditions than would simple latitude, which has typically been used for examining continental-scale gradients in parasitoid diversity (Janzen, 1981; Askew & Shaw, 1986; Gauld, 1986).

(a) Mean low temperature. This was measured as the mean temperature in the coldest month (taken to be January in the northern hemisphere and July in the southern). Five classes were distinguished: (i) 20–30 °C, (ii) 10–20 °C, (iii) 0–10 °C, (iv) –10–0 °C and (v) < –10 °C (Fig. 2.3).
(b) Range in annual temperature. This measure distinguished climates experiencing four levels of thermal variability: (i) 0–10 °C, (ii) 10–20 °C, (iii) 20–30 °C and (iv) > 30 °C (Fig. 2.4).

Both variables were generated using climatic maps from general atlases, and values were assigned for the locales where hosts were studied, based on information provided by the original workers. When a survey area encompassed more than one climatic zone, the values representing the colder and more variable climate were used. Mean low temperature and range in temperature actually represent two similar measures of local climatic conditions, since the former is the lower bound used to calculate the latter. Not surprisingly, the variables are strongly associated (Contingency coefficient = 0.773, χ^2 = 3241.0, $n = 2185$, $P << 0.001$). On the other hand, differences do exist between the measures, as, for example, when a local climate is uniformly cool (i.e. cold winters but low annual variability). Therefore, diversity patterns generated by climate could differ when using a variable incorporating annual variability from that using winter conditions, and the analyses examine both. A third climatic variable that could be important to parasitoids is the pattern and extent of rainfall, since avoidance of desiccation is likely to be critical to all insects. However, an earlier analysis suggested that annual rainfall has minimal discernible impact on parasitoid species richness patterns, except that richness is lower in deserts (Hawkins, 1990). The rainfall analysis is not repeated here.

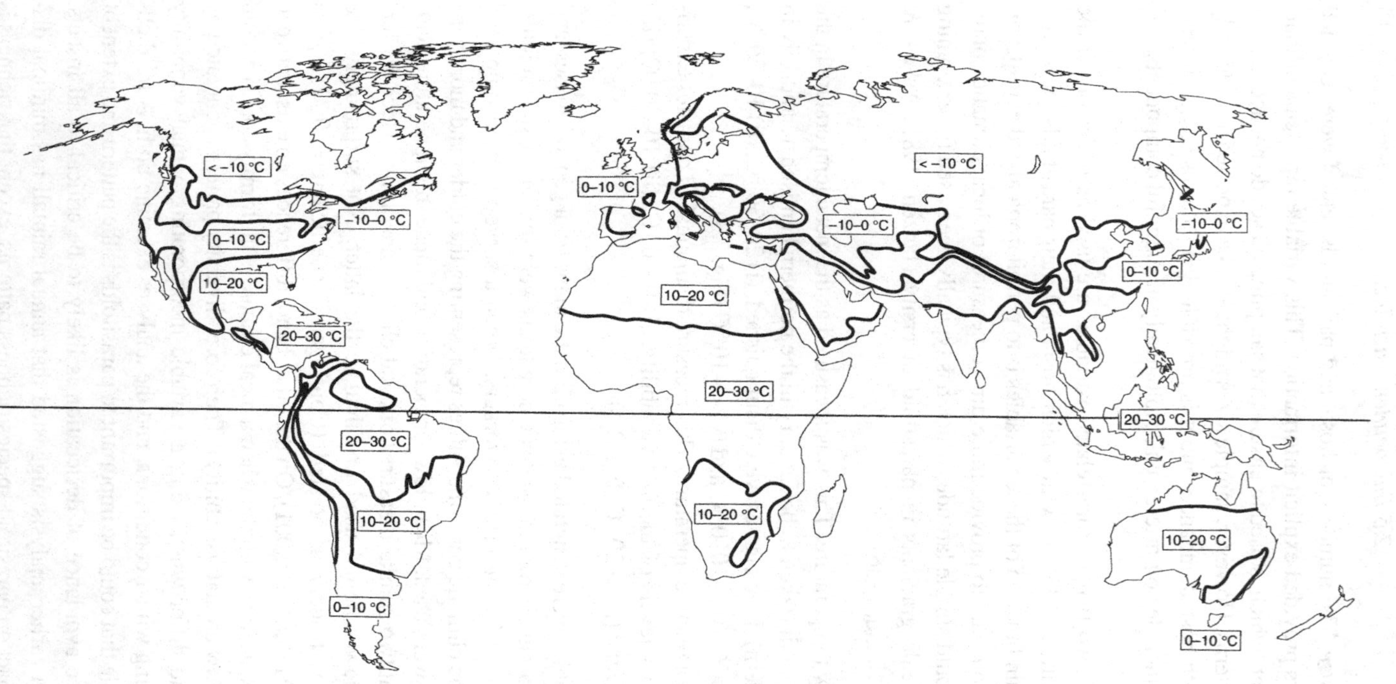

Fig. 2.3. Global distribution of mean temperatures in the coldest month.

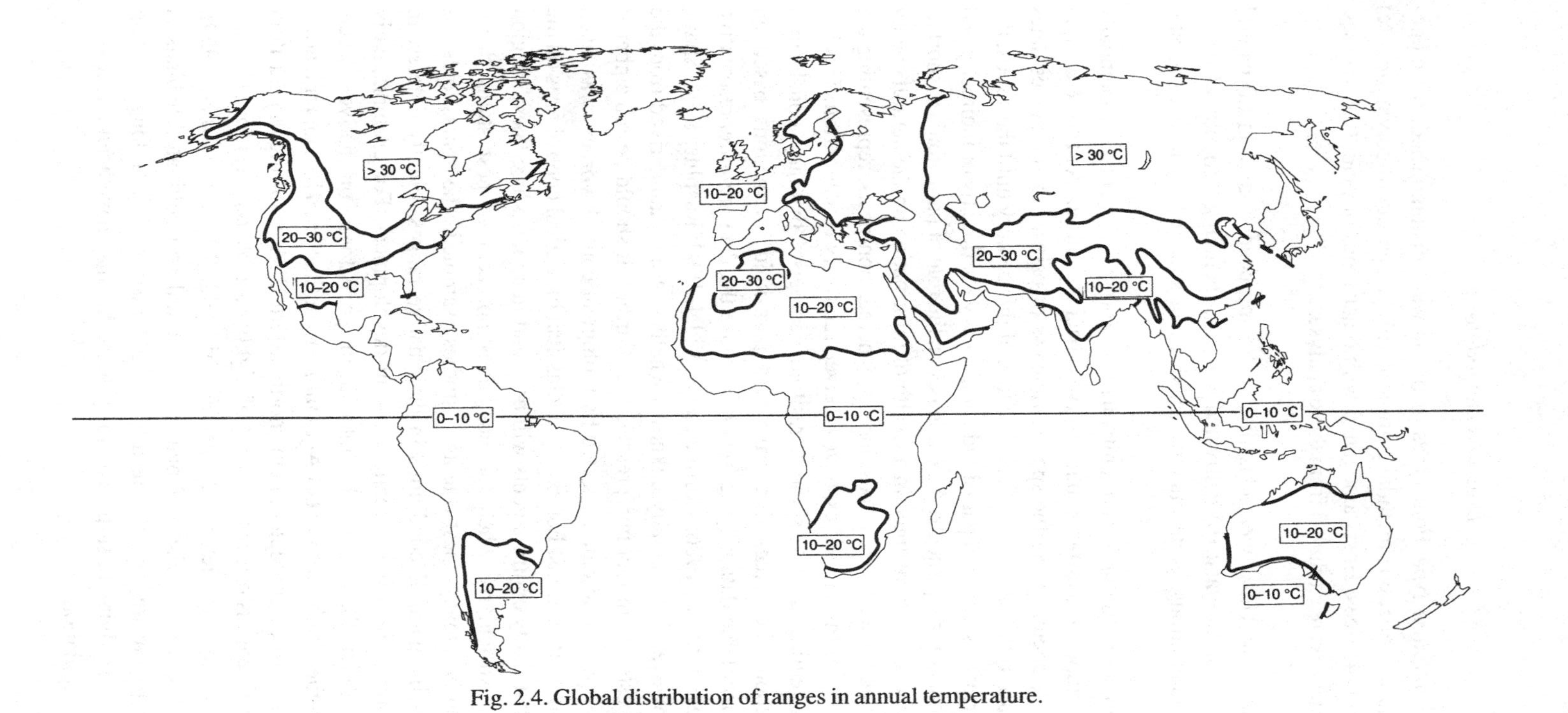

Fig. 2.4. Global distribution of ranges in annual temperature.

(6) Host foodplant type Four types of plants were distinguished: (a) monocotyledons (monocots), including grasses; (b) herbaceous dicotyledons; (c) shrubs; and (4) trees. Hosts associated with other types of plant (e.g. mosses, vines or cacti) were excluded from these analyses.

(7) Habitat type Hosts were classified as having been studied in (a) natural or semi-natural habitats; (b) cultivated habitats, including agricultural fields, orchards, plantations, plant nurseries, and urban sites; or (c) in both habitat types.

Distinguishing natural and cultivated systems represents a potentially extremely important variable that deserves special mention. As I have already pointed out, documenting the species richness of parasitoid complexes dates back at least to the late 1800s. From the turn of this century until the 1960s, the vast majority of studies on parasitoid complexes were directly or indirectly tied to biological control projects. An emphasis on the role of parasitoids as mortality agents necessarily resulted in a wealth of information on the natural enemy communities associated with pest insects, whereas non-pests, representing the vast majority of insect species, have received relatively less attention. This has lead to a feeling among some parasitoid ecologists that any conclusions reached about the forces that drive host–parasitoid interactions based on currently available data may be biased, because the data are unrepresentative (e.g. Askew & Shaw, 1986; Hoffmeister & Vidal, 1994; Rasplus, 1994; Shaw, 1994). However, it is actually unknown whether or not data derived from the study of pests are in fact unrepresentative. Further, it should be remembered that not all parasitoid data collected by biological control workers was taken in 'unnatural' conditions. For example, classical biological control rests on the premise that likely control agents will be found in an introduced pest's region of origin, particularly when the host does not achieve pest status there. Consequently, many pest-oriented studies sampled the parasitoids of herbivores in 'natural' conditions in their native ranges. Finally, interest in natural communities is not really a recent development. For example, early concerns about the impact of introduced parasitoids on native insects stimulated the US Department of Agriculture to initiate in 1915 a major rearing program of forest Lepidoptera in the northeastern USA, resulting in parasitoid community richness estimates (albeit of highly variable quality) for hundreds of non-pest host species (Schaffner & Griswold, 1934; Schaffner, 1959). Nevertheless, a comparison of patterns in natural and cultivated habitats is necessary to ensure that the results are not distorted by a long-standing emphasis on studying host–parasitoid systems that may have been affected by habitat manipulation.

2.6 Dependent variables

I used the three data sets described in Sections 2.2–2.4 to generate eight dependent variables quantifying the species richness and composition of parasitoid communities, the amount of mortality parasitoids subject their hosts to, and the ability of parasitoids to reduce host densities.

(1) Primary parasitoid species richness Defined as the total number of larval and pupal parasitoid species per host species, as described above.

(2) Parasitoid species richness by taxon Parasitoids were distinguished by order and superfamily. This permitted separate analyses of Hymenoptera and Diptera species richness. Further, among the Hymenoptera, species were distinguished as belonging to the Ichneumonoidea, Chalcidoidea, or other superfamilies, and patterns of species richness were examined for each parasitoid group.

(3) Parasitoid species richness by biology Parasitoids were classified based on the idiobiont/koinobiont dichotomy proposed by Haeselbarth (1979) and further developed by Askew & Shaw (1986). Classifications were based on the known biologies of individual species of parasitoids or by their inclusion within higher taxa with uniform biologies. Biologies are typically uniform within higher taxa of Ichneumonoidea (usually subfamily, but sometimes tribe or genus), so it was possible to categorize virtually all ichneumonoids for the entire global data set, as long as some taxonomic information was provided. For other parasitoid superfamilies, such as the Chalcidoidea, many families contain genera which include both idiobionts and koinobionts, and assignment requires detailed biological information of the species concerned. Because parasitoids were frequently not fully identified and their biologies were not studied, assignments for many species in less well studied areas were not possible. For non-ichneumonoids only parasitoids from North America and Europe were classified. Among the Diptera, all Tachinidae, Bombyliidae and Sarcophagidae were classified as koinobionts, whereas Phoridae (some of which may more properly be considered scavengers) were classed as idiobionts.

(4) Parasitoid species richness by host stage attacked Parasitoids were distinguished by whether they attack host larvae or pupae. Parasitoids attacking hosts after they have moved into concealed positions and formed cocoons (i.e. some pre-pupae and eonymphs) were classed as pupal parasitoids. Species that attack both stages were classified as larval parasitoids.

(5) Parasitoid relative abundance When workers provided host sample sizes as well as the number of individuals reared for each parasitoid species (or the number of parasitized hosts for gregarious parasitoids), species-specific parasitism rates were calculated.

(6) Hyperparasitoid species richness Defined as described above. One problem with classifying a parasitoid as a primary or a hyperparasitoid is that all facultative hyperparasitoids can be either, and the host spectrum of a parasitoid species (attacking the herbivore only, attacking both the herbivore and other primary parasitoids, or attacking only primaries) may vary among different types of host. It was decided that rather than second guessing workers, hyperparasitoids would be classified on a study by study basis, such that if a worker reported that a species had been reared solely as a hyperparasitoid in the system he was studying, I classified it as such even if it was known to be facultatively hyperparasitic in other systems. Thus, a particular parasitoid species can be classed as a primary in one study and as a hyperparasitoid in another.

(7) Parasitism rate This represents the maximum total apparent parasitism rate reported for any single host population and/or generation, although in some cases data pooled over multiple host populations/generations were used when workers did not report them separately.

(8) Biological control success rates Defined as the proportion of parasitoid introductions which have resulted in some level of pest control (whether partial, substantial or complete). The method for calculating proportions used was that referred to by Stiling (1990) as the 'total-attempts' method, in which the total number of introductions and successes were pooled without regard to the distribution of attempts against particular pest species. This is distinguished from the 'average fraction' method which weights the data by accounting for repeated introductions against particular hosts. I chose the former method for its simplicity of calculation and to maximize sample sizes. Stiling (1990) found that when comparing parasitoid establishment rates against seven independent variables, the two methods produced the same result in five cases, and he concluded that the total-attempts method did not introduce important systematic biases. Biases may be introduced when using the total attempts method when the Homoptera are included because of the extraordinary success rates for this order (Hall, Ehler & Bisabri-Ershadi, 1980), but my analyses are restricted to holometabolous hosts.

The bulk of the analyses focuses on dependent variables (1), (2), (3), (6), (7) and (8). Variables (4) and (5) were analyzed solely in support of other analyses to examine how they might influence the interpretation of results. It was found that detailed analyses of species richness by either host stage attacked or parasitoid abundance revealed nothing particularly surprising or interesting beyond what was found using other dependent variables. Given the large number of results already being reported, there is little point in presenting even more when they provide redundant results.

2.7 Statistical analysis

All statistics were generated using the SYSTAT statistical package. The parasitoid species richness data were log-transformed prior to analysis, and mortality data were angular-transformed. When transformations normalized the data sufficiently and reduced heteroscedasticity of variances, Analysis of Variance (ANOVA), simple and multiple regression, Pearson product–moment correlation, and Analysis of Covariance (ANCOVA) were used. However, it was found that in many multifactorial analyses, variances were heteroscedastistic even after transformation, so full-factorial models could not be used. Instead, the data were divided into smaller subgroups which were then analyzed separately.

In some groups of data it was found that transformation did not normalize the data sufficiently for parametric statistics. In these cases Mann–Whitney *U*-tests, Kruskal–Wallis non-parametric ANOVA and Spearman rank correlation were substituted.

The most important point concerning the analyses presented is their sheer number. Over 370 probability values are provided, and this does not include additional analyses that were conducted but are not reported. Some of these analyses can be considered *a priori*, whereas most others must be considered *a posteriori*, since previous analyses suggested that particular comparisons might be fruitful. Because my ideas on host–parasitoid interactions have been built up over the past seven years by conducting increasingly complex analyzes of larger and more detailed data sets, it is no longer possible to distinguish the *a priori* from the *a posteriori*. Therefore, the actual levels of type I error in these analyses are unknown. The main thing to remember is that it is highly dubious to consider a probability of 0.049 as significant and a probability of 0.051 as non-significant, although for practical reasons some cut-off point must be used. Some of the relationships found are relatively strong, whereas others are quite weak. The probability values can be used as a guide for distinguishing the strong from the weak, but they should not be accepted without due consideration.

Finally, because of the size and complexity of the data sets, the number of possible analyses is very large. I have not attempted to analyze all possible combinations of dependent and independent variables. Instead, I have concentrated on relationships which I consider interesting and important for understanding parasitoid communities and their interactions with their hosts. I will have undoubtedly failed to do some specific analyses that others might consider important.

3

Parasitoid species richness

3.1 Introduction

The simplest question that one can ask about a parasitoid community is, how big is it? Despite the large amount of data on the species richness of parasitoid complexes that had been accumulated during the first half of the century, it was not until the early 1960s that attention turned to explaining why some parasitoid communities were species rich, whereas others were relatively species poor. Askew's (1961) seminal work on the parasitoids of British Cynipidae was the first to compare a series of related host species to explicitly examine factors that influence the size and structure of a network of parasitoid complexes. Askew and his collaborators subsequently used this comparative approach to examine variability in parasitoid species richness and composition in cecidomyiid galls (Askew & Ruse, 1974) in a wide range of leaf mines (Askew & Shaw, 1974; Askew, 1975) and in sawflies (Askew & Shaw, 1986).

The most obvious result arising from even a cursory comparison of parasitoid complexes is that the number of parasitoids associated with individual herbivore species is highly variable. Some herbivorous insects appear to be virtually free of parasitoids (e.g. Ahmad, 1974; Janzen, 1975), whereas others support more than 50 species (e.g. Fahringer, 1941; Harman & Kulman, 1973; Delucchi, 1982). In this chapter I extend the taxonomically restricted analyses pioneered by Askew and his colleagues and examine ecological, biological and taxonomic characteristics of hosts and their foodplants that may at least partially account for this variability.

The patterns reported in this chapter form the basis for the analyses that follow in the subsequent chapters. I first examine the relationships between parasitoid species richness and the feeding niche of hosts. Askew's studies (Askew, 1975, 1980; Shaw & Askew, 1979; Askew & Shaw, 1986) have identified this as an important determinant of parasitoid community species richness, and all previous analyses of various subsets of the parasitoid assemblages represented in my data set have identified host feeding biology as the single most important correlate of how many parasitoid species a herbivore is known to support (Hawkins & Lawton, 1987, 1988; Hawkins, 1988, 1990; Hawkins, Askew & Shaw, 1990). Consequently, the effects of this variable must be taken into account in all subsequent analyses. For this to be fully justified, it is critical that the general pattern found truly reflects the influence of host feeding niche and not some hidden bias or co-factor. Therefore, I examine a number of potential biases to establish their influences on the general pattern. I then analyze other dependent variables in conjunction with host feeding niche to document additional influences on parasitoid species richness. Finally, I generate regression and ANCOVA models of increasing complexity to evaluate the relative contribution of each independent variable and to quantify the amount of variability in parasitoid species richness that can be accounted for by the available variables.

3.2 Host feeding niche

When hosts are ranked on a scale of increasing concealment within the foodplant or soil, parasitoid species richness is dome-shaped, with leaf miners and casebearers supporting the most parasitoid species and hosts both more and less exposed to the external environment supporting fewer parasitoid species (Fig. 3.1). As we shall see, this basic pattern will appear in almost all analyses involving primary parasitoid species richness, parasitoid-induced host mortality, and the impact of parasitoids on host densities, and I will address the potential mechanism(s) that might underlie this pattern at numerous points in this monograph. For now, it is most important to establish that the pattern is real and is not significantly confounded by co-factors.

There are a number of such potential confounding factors, but the most important is likely to include interactions among feeding niche and sample sizes. The number of parasitoids reared from a host species would be expected to be associated with the number of host individuals reared, and thus sample size could be an important covariable if herbivores in some feeding niches are easier to sample than those in other feeding niches. This is undoubtedly true. For example, it is usually easier to sample large numbers of individuals of a leaf miner species than of a root feeder. Therefore, it is possible that the different numbers of parasitoids from hosts in each feeding category reflect only that the parasitoid complexes of some hosts have been more intensively sampled than others, and that this is associated with feeding biology. An ANOVA of the 452 host species for which explicit sample size data are available indicates that mean sample sizes do vary among the feeding niches ($F = 2.618$, $P = 0.012$).

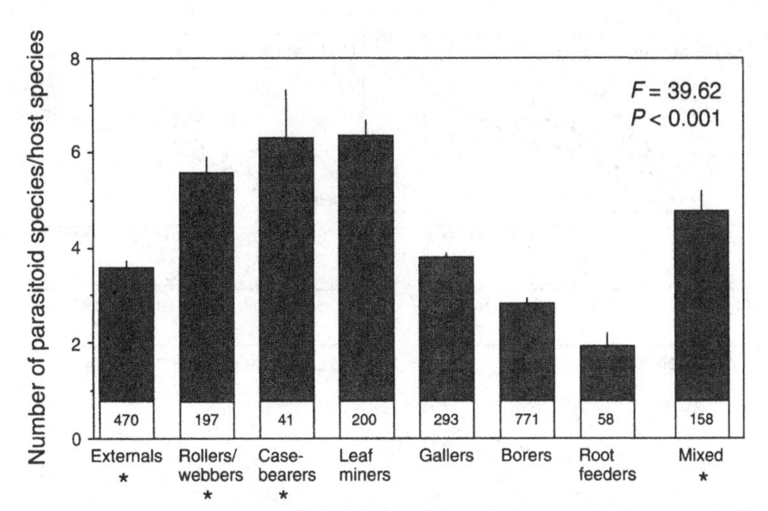

Fig. 3.1. Relationship between host feeding niche and mean parasitoid species richness. Numbers at the bases of bars are the number of host species in each niche. Vertical lines are +1 S.E.M. Asterisks under feeding niches distinguish those niches that comprise completely or partially exophytically feeding herbivores. The first seven niches are ranked in terms of decreasing mobility and increasing concealment in foodplants or in the soil. The mixed niche comprises hosts with more complicated feeding biologies which cannot be placed in the mobility/concealment ranking. Statistics from ANOVA.

Therefore, this covariable must be accounted for to properly interpret the relationship between parasitoid species richness and host feeding niche.

The influence of sample size on the pattern was examined in three ways. First, I included only those species for which workers reported rearing at least 1000 host individuals, under the assumption that a sample this large should provide a reasonably complete estimate of parasitoid assemblage size. Examining the relationships between sample size and parasitoid species richness in each feeding niche (Fig. 3.2) suggests that this may be a reasonable assumption. The vertical dashed line in each graph represents sample sizes of 1000 host individuals. If the relationships between parasitoid species richness and sample size are compared for those cases where $n < 1000$ and $n > 1000$, imaginary lines marking the upper bounds are steeply positive in most of the niches at the smaller sample sizes, indicating strong sample size effects on the maximum species richness found in each niche. In contrast, the upper bounds are flat for most niches above sample sizes of 1000. This could be interpreted to reflect that hosts in each feeding niche may support some maximum number of parasitoid species, and this maximum is detectable at sample sizes as low as 1000 hosts. Whether or not this is strictly true, it does appear that sample sizes of about 1000 delimit species richness estimates that are strongly affected by

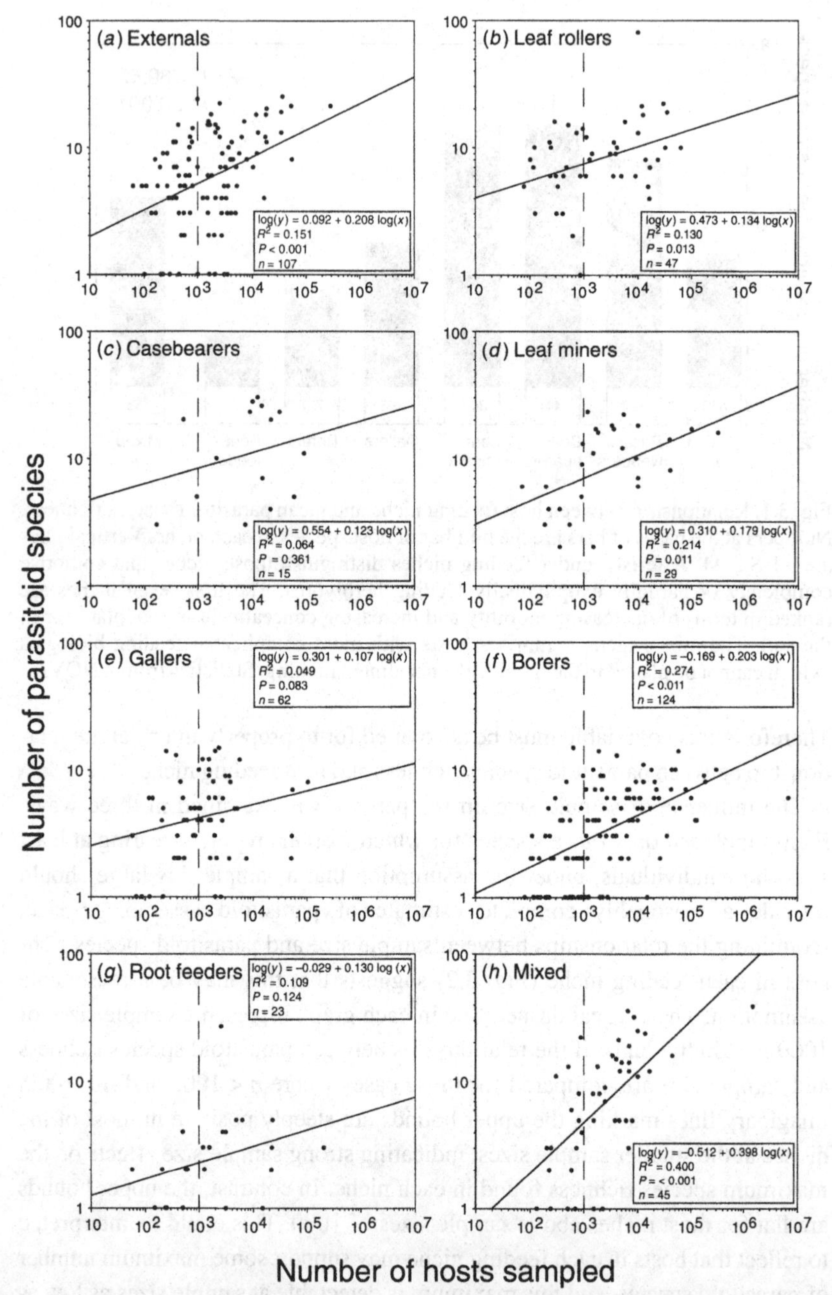

Fig. 3.2. Relationships between host sample size and parasitoid species richness for hosts in each of eight feeding niches. Vertical dashed lines delimit sample sizes of 1000 host individuals. Statistics from linear regression.

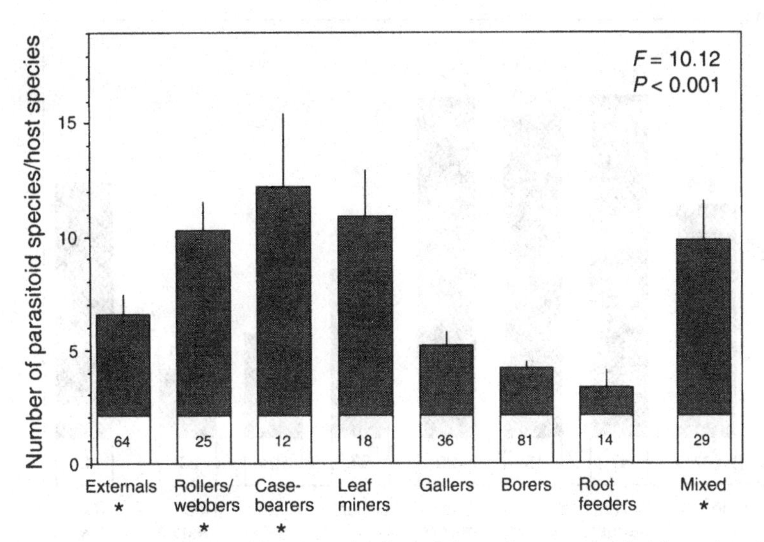

Fig. 3.3. Relationship between host feeding niche and mean parasitoid species richness for host sample sizes of ≥ 1000. Details as in Fig. 3.1. Statistics from ANOVA.

sample size and estimates that are much less affected. Therefore, analyzing only those communities based on the larger sample sizes decreases greatly the importance of sample intensity to perceived parasitoid species richness. Basing the analysis on 279 such well studied host species indicates that although parasitoid assemblages on hosts in all feeding niches are richer than indicated by the entire data set, the dome-shaped relationship is unchanged (Fig. 3.3). This represents the first line of evidence that sample size does not by itself account for the differences among the feeding niches.

Even so, sample sizes do differ among the feeding niches, even after removing less well studied systems and those for which sampling intensity is unquantified ($F = 2.287$, $P = 0.006$, ANOVA; mean sample sizes provided in Fig. 3.4). The second analysis of sample size incorporated both feeding niche and sample sizes in an ANCOVA, again using the 452 host species for which sample size data were available. The preliminary test of the feeding biology × log sample size interaction bordered on significance ($F = 1.981$, $P = 0.056$). However, this was due solely to a relatively large regression coefficient for the mixed feeding niche (Fig. 3.2). When the mixed niche was excluded, there was no heterogeneity among slopes ($F = 0.605$, $P = 0.726$). The pattern of the adjusted niche means after accounting for sample size in the ANCOVA (Fig. 3.4) appears very similar to the pattern arising from the complete data set (*cf.* Figs. 3.1, 3.4), except the leaf rollers/webbers are elevated relative to the other

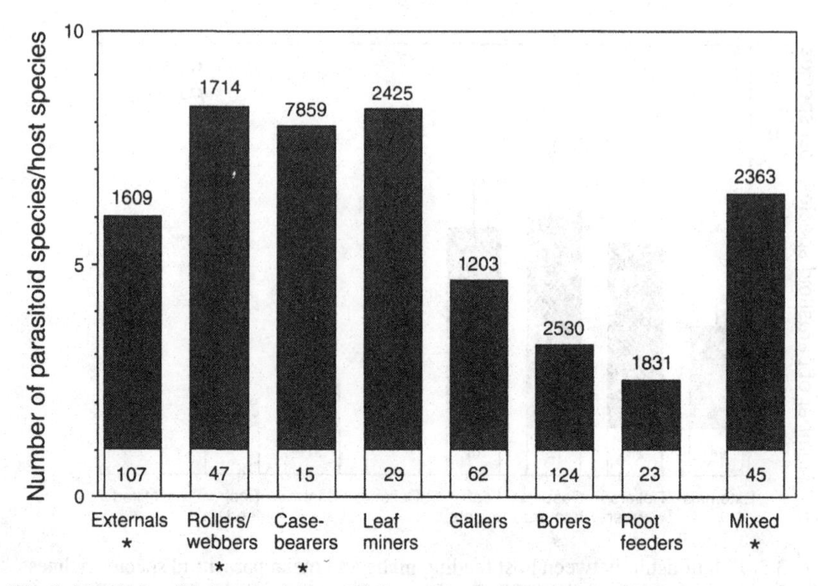

Fig. 3.4. Relationship between host feeding niche and mean parasitoid species richness, adjusted for host sample size by ANCOVA. Numbers above bars are geometric mean sample sizes for hosts in each niche; numbers at base of bars are the number of host species in each niche.

niches. But overall, the ANCOVA provides an additional line of evidence that the relative species richness of most host feeding niches do not simply reflect differences in sampling effort.

The third analysis of sampling effort assumed that the abundances of parasitoid species in an assemblage are log-normally distributed, that common species will be picked up very quickly in any sampling regimen, and increasing sample sizes will simply add increasingly rare and accidental species. Under these conditions, the numbers of the more abundant species should be more or less independent of sample size, so long as the sample sizes are greater than some bare minimum. Therefore, examining the pattern of species richness of only the relatively abundant parasitoids across feeding niches should provide an estimate of species richness which is less dependent on sampling effort.

For this analysis, I used the 253 host species for which the relative abundances of individual parasitoid species are available. Only parasitoid species achieving at least 1.0% parasitism were included. The ANCOVA of feeding niche with sample size as a covariate indicated that feeding niche was highly significant ($F = 6.481$, $P < 0.001$), whereas sample size was not ($F = 2.051$, $P = 0.153$). The resulting mean species richness figures are lower when rarer species are excluded, but the pattern of variation across the feeding niches is

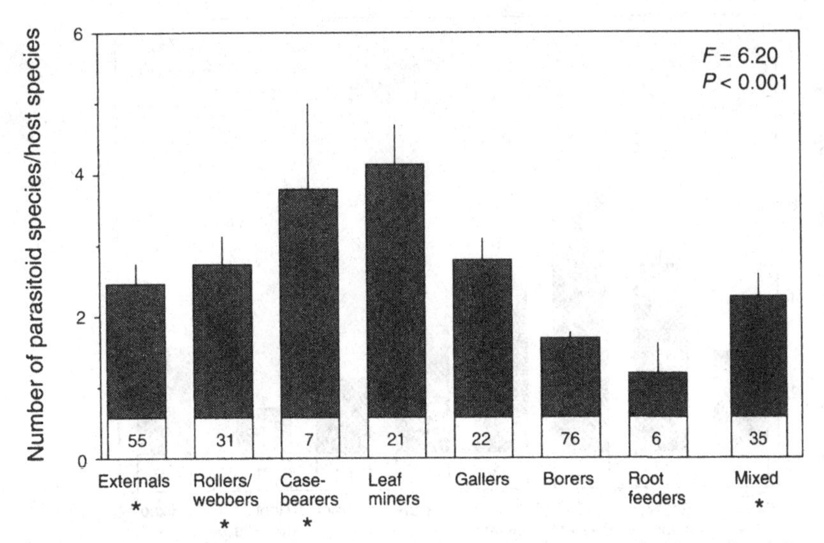

Fig. 3.5. Relationship between host feeding niche and mean parasitoid species richness, excluding parasitoids responsible for less than 1.0% parasitism. Details as in Fig. 3.1.

virtually identical to that for all parasitoid species (Fig. 3.5). Therefore, excluding rarer parasitoids and decoupling the pattern of species richness from sampling intensity gives a qualitatively similar picture of the influence of host feeding niche on parasitoid species richness patterns. This analysis, and the preceding two, make it clear that the feeding niche pattern is quite robust with regard to sampling intensity, and that feeding niche has a strong independent effect on parasitoid species richness irrespective of how well known the parasitoid complex is.

A second potential co-factor that could influence the pattern arises from an interaction between host taxonomy and feeding biologies. Herbivore taxa are not randomly distributed among the feeding niches, and if evolutionary history is a more important determinant of parasitoid species richness than are ecological characteristics of hosts, including a wide range of host taxa in the analysis could affect parasitoid species richness patterns in a number of ways. First, including more than one species of a restricted host taxon (at say the genus or family level) may essentially represent pseudoreplication (Harvey & Pagel, 1991), since the parasitoid assemblages associated with closely related host taxa are likely to include many of the same parasitoid species. If these host taxa also all have similar feeding biologies, this could distort the patterns of parasitoid species richness. An example of the inclusion of many related taxa with shared parasitoid species occurs among the leaf miners, in which 20 (10%) of

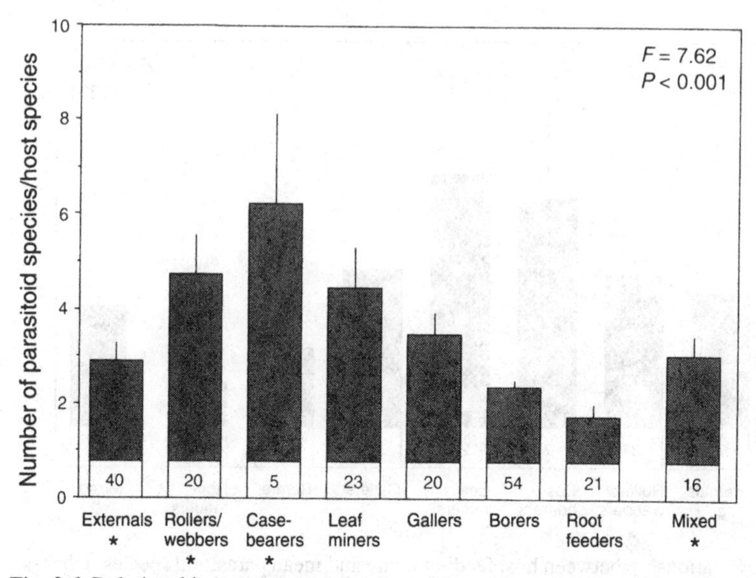

Fig. 3.6. Relationship between host feeding niche and mean parasitoid species richness, using mean parasitoid species richness per host species per host family in each niche. Hence, numbers at bases of bars are the number of host families in each niche.

the 200 species in the global data set are *Phyllonoryctor* spp. (Lepidoptera: Gracillariidae). This group has been well studied in England (Miller, 1973; Askew & Shaw, 1974; Shaw & Askew, 1976a; Askew, 1980), and the parasitoids have been fully identified, making it possible to determine the amount of overlap among assemblages. The 280 host–parasitoid records in Britain actually represent only 55 parasitoid species. Therefore, 80.4% of the species richness data in this host genus are multiple records. This kind of replication is likely to occur in all feeding niches where several congenerics or confamilials are included, although because many parasitoids are not fully identified it is not possible to measure the extent. The question here is how does this affect the apparent species richness pattern found among the host feeding niches?

To examine the effect of including multiple cases of related host taxa within individual feeding niches, I calculated the mean number of parasitoid species per host species for all hosts in each host family in each feeding category. Thus, each host family is represented by a single value in each feeding niche, eliminating the possible influence of pseudoreplication at the host family level. The resulting differences in mean species richness remain highly statistically different (Fig. 3.6), but casebearers have greater mean assemblage size than leaf miners (but not significantly so; $P = 0.298$, post-hoc contrast). Despite this

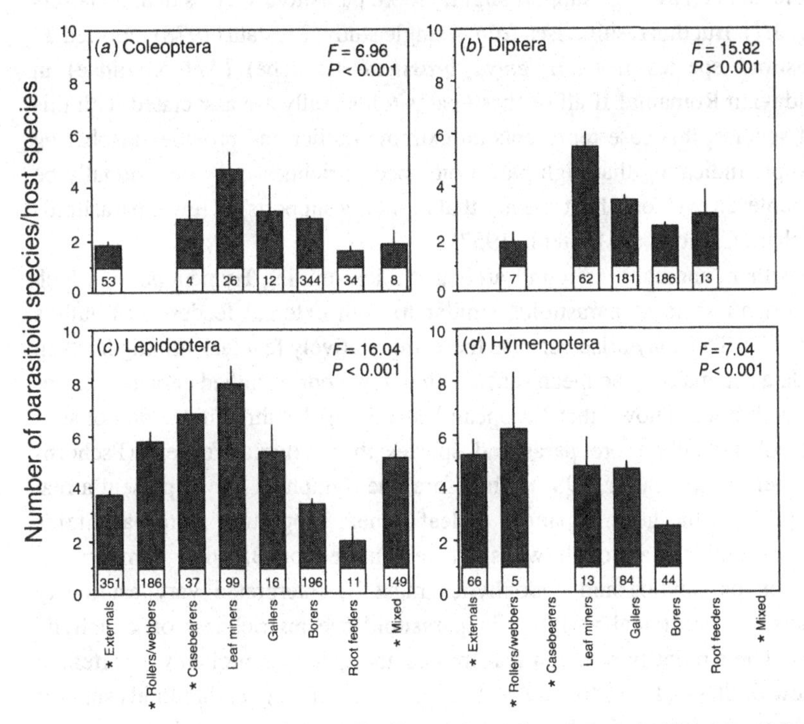

Fig. 3.7. Relationship between host feeding niche and mean parasitoid species richness for hosts in each of four orders. Details as in Fig. 3.1. Statistics from ANOVA.

small difference in the pattern, the general shape of the relationship between species richness and the feeding niches is very similar to that found when all host species are included individually (*cf.* Figs. 3.1, 3.6). Therefore, as with sample size, the pattern is reasonably robust to this potential taxonomic co-factor.

A second way that host taxonomy may affect the general pattern of species richness among host feeding niches is that most host taxa are restricted to a limited range of feeding niches. For example, there are very few exophytic Diptera, whereas there are many that are leaf miners and gallers. It is known that parasitoid assemblage sizes vary among host orders in Britain (Hawkins & Lawton, 1987), so if each order is differentially represented among the feeding niches, taxonomically based differences in the means of niches could be substantial. Examining each host order separately indicates that feeding niche significantly influences parasitoid species richness in all four cases (Fig. 3.7). Furthermore, all tend to show the dome-shaped relationship found in the general analysis. However, apparent exceptions do exist. For example, dipterous root feeders on average support slightly more parasitoid species than do borers (Fig. 3.7). But this result arises from a single study: Mustata (1978) reported 27 parasitoid species from *Hylemya brassicae* (Bouché) (Anthomyiidae) in Moldavian Romania! If all of these parasitoids really are associated with this host species, this case represents an extreme outlier and provides a sobering example indicating that high parasitoid species richness may occasionally be encountered even on a host species that normally supports far fewer parasitoids (Wishart, Colhoun & Monteith, 1957).

Another 'anomaly' is found among the leaf-mining Hymenoptera, which support numbers of parasitoids similar to both external feeders and gallers (Fig. 3.7). This may arise because there are relatively few leaf-mining hosts in the data set, making the mean sensitive to a few poorly studied parasitoid complexes, but it is known that European leaf-mining Tenthredinidae do not support substantially more parasitoid species than external feeders (Pschorn-Walcher & Altenhofer, 1989). Therefore, the Symphyta may represent a real exception to the general pattern of leaf miners supporting more parasitoids than exophytics; although we shall see in Section 3.4 that comparisons between exophytics and endophytics must be interpreted very cautiously because of latitudinal variation in parasitoid species richness on exophytic hosts. The similarity between leaf miners and gallers is probably an artefact: Askew & Shaw (1986) found that European leaf-mining tenthredinids support twice as many parasitoids as gallers on average (12.1 vs 6.5). Therefore, despite the odd exceptions, for the bulk of the data the relationship between parasitoid species richness and feeding niche appears quite robust at the host-order level of taxonomic discrimination; hosts in all four orders contribute to the general shape of the pattern, and including all host orders in analyses, while undoubtedly adding scatter to the data, does not seriously bias resultant patterns.

More subtle evolutionary effects are also possible. Godfray (1993) has hypothesized that host taxonomic homogeneity may account for differences among feeding niches. He points out that many hosts belong to genera with large numbers of ecologically similar species, creating a 'swarm of species'. This swarm should be a larger evolutionary target for parasitoids than a host belonging to a taxonomically restricted group. Generalist parasitoids that recruit into the swarm will be able to attack most or all of its members, elevating parasitoid richness on all host species. It should also be easier for specialists to shift hosts if the latter are closely related, enhancing the species richness of parasitoid complexes associated with individual host species. This reasoning also forms the basis for the potential importance of taxonomic 'pseudoreplication' tested above.

If host taxonomic homogeneity is not randomly distributed among the host species in the data set, and species swarms tend to occur in some feeding niches and not in others, this could give rise to the niche differences. To determine if a taxonomic homogeneity × feeding niche interaction exists, I calculated the mean number of parasitoid species per host genus in Great Britain, the only country where herbivore generic diversity has been reasonably well established (Section 2.5). These generic means can be used to test the effects of feeding niche and taxonomic homogeneity separately and in concert. Using this much more restricted data set, feeding niche is highly significant ($F = 4.93$, $P < 0.001$, $R^2 = 0.240$, $n = 84$). Host generic diversity (log-transformed) is also significantly positively associated with mean parasitoid species richness ($F = 8.46$, $P = 0.005$, $R^2 = 0.094$) but not as strongly as is feeding niche. Importantly, when feeding niche is analyzed using ANCOVA with generic diversity as the covariate, the niche × generic diversity interaction is not significant ($F = 1.11$, $P = 0.362$). The final ANCOVA (with the interaction term removed) indicates that both factors are significant (feeding niche $F = 4.62$, $P = 0.001$; generic diversity $F = 6.92$, $P = 0.010$; multiple $R^2 = 0.303$) but, as when they are analyzed separately, the result for feeding niche is much stronger (the sum of squares for feeding niche was 3.3 times as large as that for generic diversity). Godfray's (1993) prediction that richer parasitoid complexes are associated with hosts belonging to more species-rich genera is supported within the host feeding niches, but the effects of taxonomic richness and feeding niche are independent (the interaction term is non-significant), and taxonomic diversity cannot explain differences between feeding niches, at least in Britain.

A final potential complicating factor in the analysis of host feeding niche is that hosts were assigned based on habits of larvae, but the parasitoid data include both larval and pupal parasitoids. Pupal parasitoids, by definition, attack the host after it has stopped feeding and chosen a pupation site, which in many cases will not be in the larval feeding site. Do larval and pupal parasitoid species richness levels vary among the feeding niches similarly, or do interactions exist which may produce a biased picture of species richness when hosts are classified on the basis of larval habits only?

The species richness patterns of those parasitoids which attack the host as larvae in their feeding site show a dome-shaped relationship with feeding niche that is very similar to that when all parasitoids are included (Fig. 3.8). There are two points concerning the pupal parasitoids that are relevant. First, including pupal parasitoids has minimal effect on the overall pattern, largely because they contribute only a small proportion to the total species richness of the complexes. This almost certainly represents an underestimation of the true richness of pupal parasitoid species relative to larval parasitoids, since whereas larvae

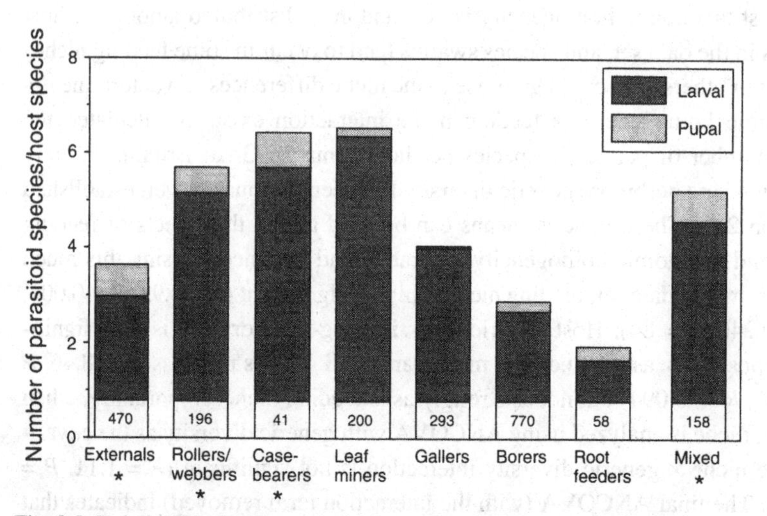

Fig. 3.8. Relationship between host feeding niche and mean parasitoid species richness of larvae-attacking and pupae-attacking parasitoids. Details as in Fig. 3.1.

have been reared for the vast majority of host species in the data set, pupae have been sampled for far fewer. But as far as the analysis of feeding niches is concerned, it makes little difference to the interpretation of the patterns whether pupal parasitoids are included or not.

Second, the pattern for pupal parasitoids does not simply follow that of the larval parasitoids, although the host feeding niches support substantially different numbers of pupae ($\chi^2 = 229.14$, $P < 0.001$, Kruskal–Wallis ANOVA). Pupal parasitoid species richness tends to be higher in feeding niches representing hosts that are completely or partially exophytic and is lower in hosts feeding completely endophytically (Fig. 3.8). This almost certainly reflects the relative richness of idiobionts and koinobionts (see Chapter 4). Hosts in all feeding niches support both types of parasitoids; in the case of exophytics, all koinobionts attack larval stages, whereas virtually all idiobionts are restricted to pre-pupae or pupae that are attacked after the host has concealed itself. On endophytics, idiobionts need not wait until the host has concealed itself for pupation to attack them and, consequently, many attack larvae and pupae indiscriminately. Therefore, parasitoid communities centered on exophytic hosts are neatly divided into two discrete components based on host stage and parasitoid developmental syndrome, whereas those on endophytic hosts are less so, with many 'pupal' parasitoids attacking host larvae at least some of the time. I will return to more detailed treatments of the koinobiont and idiobiont components of parasitoid communities in the next chapter.

All of the above analyzes indicate that the relationship between parasitoid species richness and the feeding niches of hosts is only marginally influenced by the major potential biases that we might expect. Therefore, the available evidence indicates that the pattern is real and robust, and host feeding niche can be used as a template against which to examine other forces that influence parasitoid community size and structure.

3.3 Foodplant/habitat

The type of plant on which a herbivore feeds has been shown to influence the species richness of associated parasitoid complexes for some herbivore groups, including Cynipidae and *Phyllonoryctor* (Gracillariidae) (Askew, 1980), *Coleophora* (Coleophoridae) (Lampe, 1984), seed-feeding beetles of African legumes (Rasplus, 1994), and British Lepidoptera and Hymenoptera in general (Hawkins, 1988). All of these results indicate that parasitoid species richness is lowest for hosts feeding on monocots and/or herbs and greatest for hosts feeding on trees. On the other hand, for other herbivore groups the pattern of increasing parasitoid species richness along the series monocots—herbs—shrubs—trees has not been found, for example, Agromyzidae (Askew, 1980; Hawkins, 1988), Homoptera (Hawkins, 1988), Cecidomyiidae (Hawkins & Gagné, 1989) and Tortricoidea (Mills, 1993). Further, Hawkins & Lawton (1987) found for British herbivores that parasitoid species richness was affected by foodplant type when this factor was examined in isolation but was not when host feeding niche was included as a co-factor in a multiple regression. A subsequent analysis of British holometabolous herbivores indicated that parasitoid richness was greater on trees for exophytics (externals and rollers/webbers), leaf miners and gallers but not for borers (Hawkins *et al.*, 1990). The range of results found so far raise the question of how general the effects of host foodplants are on the sizes of parasitoid communities and suggest that several factors may be interacting to produce the patterns for specific groups of herbivores.

Foodplant type does not represent a single, simple factor to which insects may respond. Herbs and trees, for example, may simultaneously differ in their size, growth form, seasonal development, the variety and persistence of individual parts (Lawton, 1983), or their chemistry (Feeny, 1976). All of these components reflect a plant's overall 'architecture' or 'apparency' and can be expected to influence both herbivore and parasitoid communities individually or in concert. To complicate the issue even further, plant architecture is compounded with other ecologically relevant factors. In natural habitats, trees invariably occur late in the successional cycle, whereas herbs more often col-

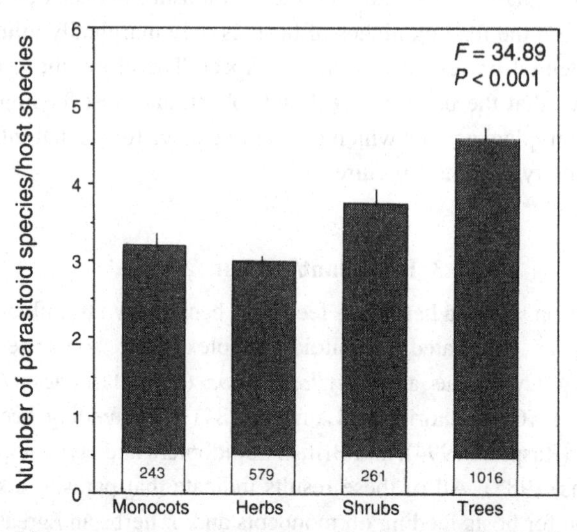

Fig. 3.9. Mean parasitoid species richness for hosts feeding on four types of plants varying in size and/or architectural complexity. Details as in Fig.3.1. Statistics from ANOVA.

onize disturbed habitats in early stages of succession. Properties of plant communities generated by succession may have profound influences on herbivores and parasitoids in ways not related just to plant architecture (Price, 1991). Finally, a plant can occupy a habitat created by man. A soybean field and a two-year-old patch recovering from a forest fire are hardly equivalent, despite both being dominated by herbs, and habitat disruption has been shown to influence parasitoid community size and structure in at least one instance (Miller, 1980; Miller & Ehler, 1990). For all of these reasons, attempts to determine and explain the effects of host foodplant type on parasitoid assemblage size must be fraught with complications, and it is difficult to judge whether differences among plant types found so far reflect the effects of inherent characteristics of the plants themselves or those of factors associated with the habitats in which the plants occur.

In this section I examine the general relationship between parasitoid species richness and the host foodplant type. However, because characteristics of each type of plant are confounded with habitat characteristics, I also compare natural and cultivated habitats in an attempt to tease out possible interactions with plant type.

When all hosts are compared with regard to plant type only, there is an increase in parasitoid species richness with increasingly complex plant architecture (monocots = herbs < shrubs < trees) (Fig. 3.9). Further, the relationship

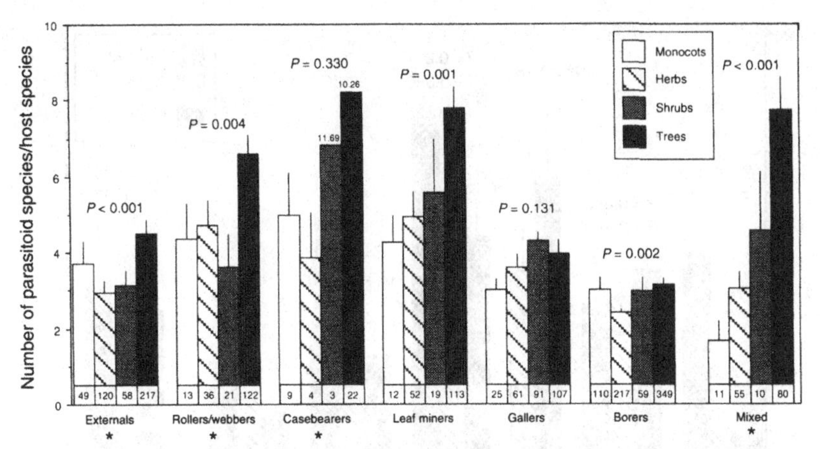

Fig. 3.10. Mean parasitoid species richness for hosts in each of seven feeding niches (root feeders excluded) on four plant types. Probabilities are from single-classification ANOVA conducted on hosts in each niche. Smaller numbers above casebearers are +1 S.E.M., which extend beyond the scale of the figure. Details as in Fig. 3.1.

appears to broadly hold across some, but not all, host feeding niches (Fig. 3.10) (root feeders have been excluded from these analyzes). Galler and casebearer parasitoids do not show a significant relationship at all, and although borer parasitoids differ significantly, they do not show a consistent increase with plant types of increasingly complex architecture. For externals and leaf rollers/webbers, parasitoid richness is highest on trees, but the progression from monocots to shrubs is not consistent. It is only among leaf-miner and mixed exo-/endophytic parasitoids that a consistent increase with increasingly complex plant architecture is observed (Fig. 3.10). Despite the variable results, it is notable that parasitoid species richness is highest on trees in six of the seven niches.

Before examining plant type in more detail, it is necessary to compare natural and man-made habitats in order to determine how human manipulation affects parasitoid communities. Of the 1868 hosts for which it was possible to classify the type of habitat in which the parasitoid complex was sampled, 899 (48.1%) were studied in natural or semi-natural habitats, 820 (43.9%) were studied in cultivated habitats, and 149 (8.0%) were studied in both. Thus, the plant type results could be affected by the inclusion of large numbers of cases where communities are to some extent 'unnatural'.

Despite the obvious and important ecological differences between natural and cultivated habitats, hosts occupying these habitats support surprisingly similarly sized parasitoid complexes (Fig. 3.11). When natural and cultivated habitats are compared directly, the only significant difference is found for

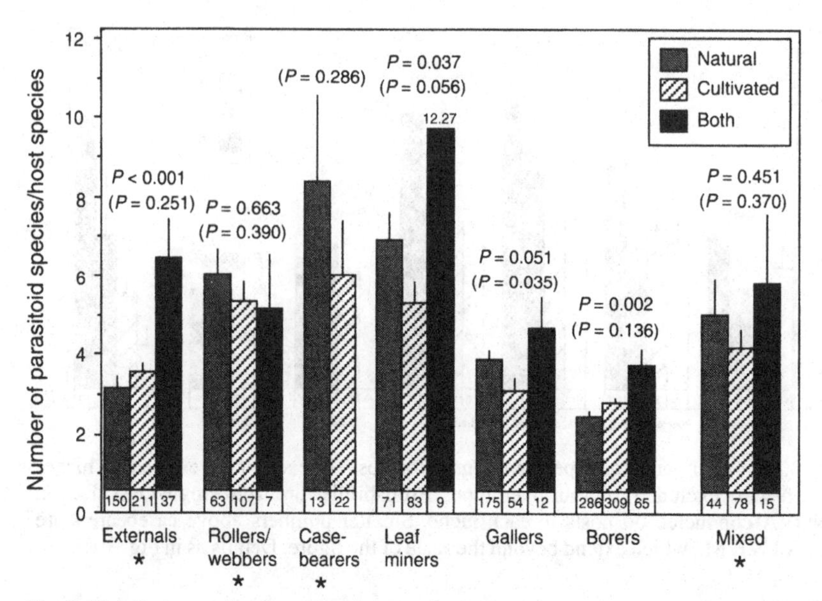

Fig. 3.11. Mean parasitoid species richness for hosts in each of seven feeding niches for hosts studied in either natural or cultivated habitats, or both. Parenthetic probabilities arise from comparisons of natural vs cultivated habitats only; other probabilities compare all three habitat types. Details as in Fig. 3.1. Smaller numbers above leaf miners are +1 S.E.M., which extend beyond the scale of the figure.

gallers, and even in that case the difference is weak (Fig. 3.11, probabilities in parentheses). In general, habitat manipulation has minimal apparent influence on parasitoid species richness. On the other hand, when hosts that have been studied in both natural and cultivated habitats are included, parasitoid species richness tends to be higher than in either natural or cultivated habitats considered separately (Fig. 3.11). If this does not arise from differences in sampling intensity, it suggests that habitat manipulation may often influence the species composition of the parasitoid complex, with one set of parasitoids attacking hosts in their natural habitat, and another set attacking them in the cultivated habitat. If so, this is an important effect, but it does not alter the conclusion that, in terms of total species richness, hosts in natural and manipulated habitats generally support very similarly sized parasitoid complexes.

It appears that plant type does tend to influence parasitoid assemblage size, but that habitat type does not. But what of interactions between the two? A positive association between plant type and parasitoid species richness has been generally explained by the high apparency of trees, which results in more abundant, diverse, and predictable herbivore populations. These enhanced resources then provide the template for richer parasitoid communities (Askew & Shaw, 1986). Further, Price (1991) has identified a wide range of habitat, plant and herbivore characteristics associated with ecological succession that are predicted to influence the sizes of parasitoid complexes. Since herbs generally dominate early successional stages, whereas trees compose the dominant plants of later succession (except of course in grasslands, marshes, etc.), insect communities inhabiting natural habitats will be influenced by covarying plant type and habitat factors simultaneously. Much of this covariation can be partitioned by comparing natural versus cultivated habitats, because cultivation decouples some (but not all) of the plant and habitat traits associated with succession. For example, herbivore populations in early succession tend to be relatively small, since plant patches are themselves small; field crops, in contrast, are often grown in large patches and their herbivore populations may be huge. Also, early successional plants may utilize toxic defences which affect both herbivores and parasitoids, unlike most crop plants. Finally, although plants grown as field crops share with natural herbs the trait of tending to be short-lived as individuals, similar agricultural plant 'communities' may persist for long periods of time at the same site if crops are not rotated. This persistence of resources in time will be enhanced even more when large contiguous areas are devoted to agriculture, and all but the most immobile insects can move between fields growing the same crop plants in different years.

At the other end of the spectrum, a natural forest offers a diverse array of potential hosts to generalist parasitoids, with each tree species supporting scores to hundreds of herbivore species. A tree plantation or orchard, in contrast, will support fewer such potential host species. Further, compared to a natural forest, patch size in orchards and many tree plantations may be small, sometimes smaller than that for field crops.

How does this wide range of possible plant–habitat interactions influence parasitoid community richness, and to what extent does the inclusion of parasitoid data from both natural and cultivated habitats obscure plant type effects? The patterns among the four types of plant studied in natural habitats appear very similar to those found in the previous analysis which included all herbivores (Fig. 3.12*a*). Parasitoid species richness is highest on trees on all three types of exophytic host and leaf miners but is not on gallers and borers (*cf.* Fig. 3.10). The only real difference is that the pattern for external feeders is much clearer in natural habitats, with a consistent rise in parasitoid species richness with increasingly complex plant architecture.

In contrast, plant effects appear very weak in agricultural habitats, and a significant gradient is found only for the mixed exo-/endophytics (Fig. 3.12*b*). Thus, there are minimal apparent differences among plant types in cultivated

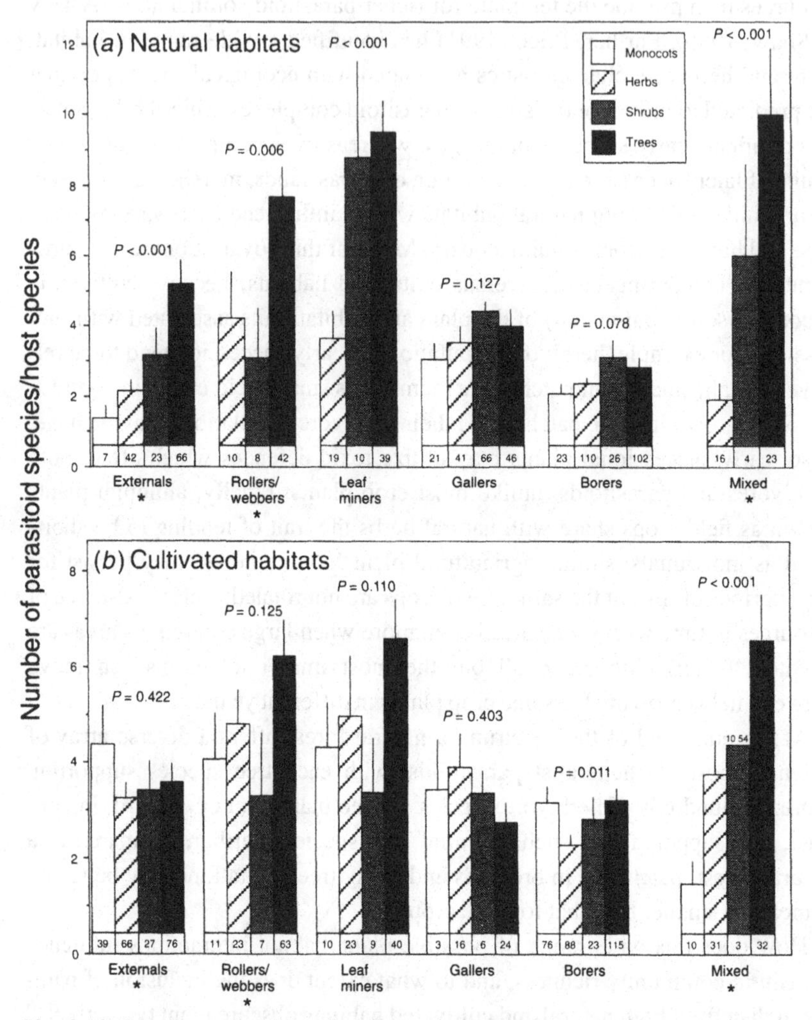

Fig. 3.12. Mean parasitoid species richness (+1 S.E.M.) by host foodplant type in (*a*) natural and (*b*) cultivated habitats. Details as in Fig. 3.1.

habitats, whereas at least some feeding niches show a relationship with plant type in natural habitats. This suggests that the effects of plant type on parasitoid species richness arise largely from characteristics of the habitats in which the plants grow (e.g. plant and/or herbivore diversity, patch size and habitat stability) rather than from innate characteristics of individual plants (e.g. size, architectural complexity, longevity).

It is noteworthy that galler parasitoids do not show a relationship with plant

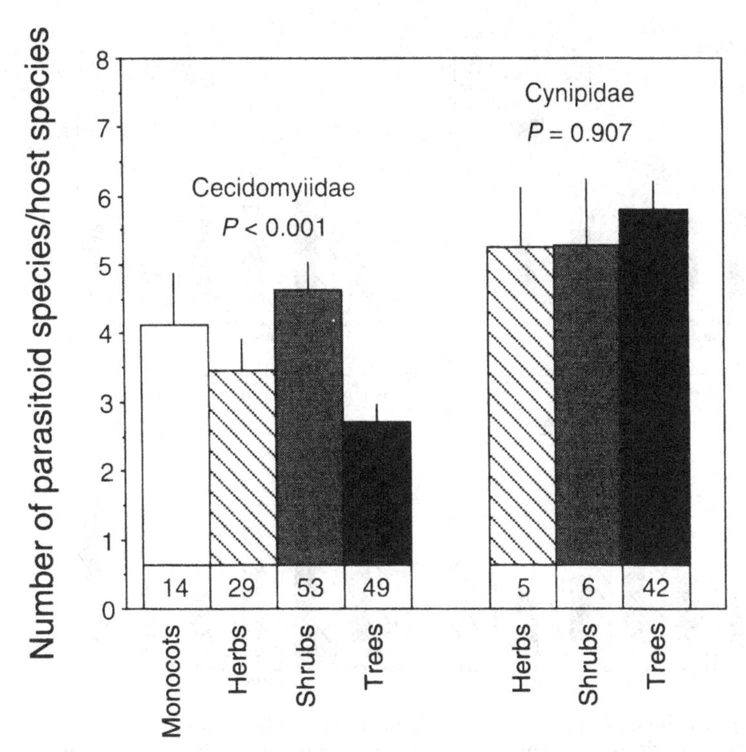

Fig. 3.13. Relationships between mean parasitoid species richness (+1 S.E.M.) and plant type for gall-forming Cecidomyiidae (Diptera) and Cynipidae (Hymenoptera). Numbers are the number of host species. Statistics from ANOVA.

type in either natural or cultivated habitats in the global data, since positive relationships have already been found for cynipids (Askew, 1980), as well as for gallers in general in Britain (Hawkins *et al.*, 1990). It could be that strong differences between the two taxonomic groups that dominate the galler niche explain at least part of this result; 69.7% of the galler species comprise Cynipidae and Cecidomyiidae. Cynipid parasitoid species richness is highest on hosts on trees, at least in Britain (Askew, 1980), whereas a global sample of cecidomyiids has indicated that parasitoid species richness is highest on monocots and shrubs and lowest on trees (Hawkins & Gagné, 1989). Examining these galler groups separately, the pattern for cecidomyiids reported by Hawkins & Gagné (1989) remains similar for the data analyzed here (Fig. 3.13), but further, cynipids do not differ among the plant types. It appears that the positive association of parasitoid species richness and plant type found for Britain does not hold (or is not detectable) in a global sample of gallers, and

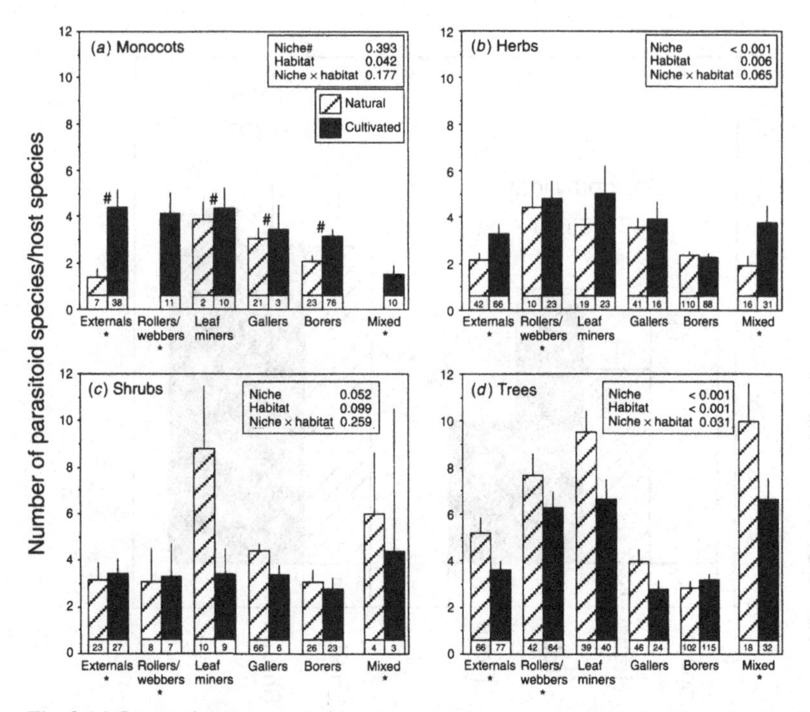

Fig. 3.14. Interactions among habitat type and plant type for hosts in six feeding niches. Parasitoid species richness in natural and cultivated habitats for hosts on (*a*) monocots, (*b*) herbs, (*c*) shrubs, and (*d*) trees. Rollers/webbers and mixed feeders were excluded from the analysis of monocots. Numbers in the boxes represent probability values for main effects and interactions from two-way ANOVAs.

galler parasitoid species richness may not increase with increasingly complex plant architecture everywhere.

Another way of viewing the interactions among plants and habitats is to directly compare the parasitoid complexes associated with each herbivore feeding niche in natural versus cultivated habitats for each plant type. Although habitat type in general appears to have minimal effects on parasitoid species richness (Fig. 3.11), it is possible that specific combinations of natural versus cultivated habitats and the four plant types may. Indeed, such complex interactions are present (Fig. 3.14). For the herbivores associated with monocots and herbs, parasitoid complexes are significantly richer in cultivated habitats (Fig. 3.14*a*,*b*). On shrubs, greater richness starts to shift towards natural habitats, with the overall difference between the habitats no longer significant (Fig. 3.14*c*). Finally, on trees, parasitoid richness is greater in natural habitats in five of the six feeding niches analyzed (Fig. 3.14*d*).

Again, habitat characteristics appear to be as important as plant characteristics in influencing associated parasitoid communities. Hosts on trees do not support inordinately rich parasitoid complexes in orchards, plantations or nurseries. It is only in natural habitats (i.e. in late succession) where parasitoid richness is high.

The conversion of natural habitats into cultivated ones typically modifies at least two factors known to influence parasitoids, but the outcome of these modifications might be expected to produce opposing patterns (see Sheehan (1986) for discussion of the related topic of agroecosystem diversification and natural enemies). First, cultivation reduces plant diversity, which will consequently reduce herbivore diversity. If generalist parasitoids are attracted to plants/habitats with larger numbers of potential host species (Askew, 1980), parasitoid species richness should be lower in simplified habitats. But second, the concentration of plant resources associated with cultivation will simultaneously increase the local abundances of those herbivores that can use the plant being cultivated (except perhaps in the face of continuous insecticidal control). All else being equal, abundant hosts should support larger parasitoid complexes than rare hosts, because specialists should be able to maintain larger populations on abundant hosts, reducing the chances of local extinction, and generalists should encounter abundant hosts more often. Therefore, habitat simplification could actually increase parasitoid species richness. The question is, which of these effects is generally the more important?

Comparing the more extensive data from herbs and trees suggests that both factors are operating, but in different habitat extremes (Fig. 3.14). The richer parasitoid complexes on herbs in cultivated habitats suggests that the overall effect of concentrating host resources gains more parasitoids than are lost by losses in plant/herbivore diversity, whereas the reduction in parasitoids on trees in cultivated habitats suggests that reducing plant/herbivore diversity has a more powerful effect than the concentration of hosts on one to a few tree species. This interaction between plant type and habitat type, with a discontinuity of effects of habitat simplification on plants generally composing disturbed habitats and on those composing stable habitats, yet again implicates the properties of habitats as being at least as important as the properties of individual plant species on parasitoid community richness. Parasitoid communities on trees are diverse not simply because their hosts are on larger, longer lived plants, but also because the trees are part of diverse plant communities. On the other hand, inherent plant properties cannot be entirely dismissed. Even in cultivated habitats, where parasitoid richness on herbs is highest and richness on trees is lowest, hosts on trees still support significantly richer parasitoid complexes (two-way ANOVA of niche and plant; $F = 14.051$, $P < 0.001$, and $F =$

5.797, $P = 0.016$, respectively; interaction $F = 1.732$, $P = 0.125$). Clearly, there are a number of subtle interactions that complicate our understanding of the tri-trophic level interactions among plants, herbivores and parasitoids.

3.4 Latitude/climate

The above factors would be expected to operate locally, or at least over fairly small spatial scales. But what of larger scale effects? Forest leaf miners and casebearers support the largest parasitoid complexes overall, but is this equally true in both Britain and Brazil? Latitudinal patterns in total parasitoid species richness have been reasonably well established for at least some parasitoid groups. Owen & Owen (1974) trapped flying ichneumonids in four gardens in Britain, Sweden, Kenya, and Sierra Leone and found that they were no more diverse in the two tropical sites. Subsequently, Janzen & Pond (1975) found more species of parasitic Hymenoptera (including Ichneumonidae) in sweep samples from an English old field than from secondary vegetation in Costa Rica. Janzen (1981) also reported that North American ichneumonids were most diverse between 38° N and 42° N, whose mid-range runs through the central USA from approximately New Jersey to northern California. Similarly, Gauld (1986) found that the majority of ichneumonid subfamilies are richer in extra-tropical Australia than tropical Australia, and total richness was slightly lower in the tropics. Finally, Askew (1990) found that malaise-trapped ichneumonids were less diverse in Sulawesi (Indonesia) than in Britain or France.

All available data indicate that for the largest family of parasitoids, the Ichneumonidae, the latitudinal 'gradient' of species richness is either flat or declines slightly towards the tropics. Latitudinal patterns for other groups of parasitoids are less well documented, but it appears that for some families species richness is highest in the tropics. Noyes (1989) and Askew (1990) compared the diversity of parasitoids from Indonesian samples with European diversities and concluded that for many chalcidoid families, for example, tropical species richness is greater. Hespenheide (1979) collected more than three times as many Chalcididae in Panama and Costa Rica in a year than had been described for the entirety of North America. It appears there is no single latitudinal species richness gradient for all parasitoids; some families are richer in the tropics, some are not.

All of these gradients, in so far as they are known, relate to the absolute, regional parasitoid species richness. But an ecologically more relevant question is, how does parasitoid species richness vary latitudinally in relation to their hosts? The creation of new species is an evolutionary matter, but their maintenance once they have evolved depends on ecological processes. Species exist in ecological time inextricably tied to their hosts as parts of parasitoid complexes, and total parasitoid species richness can be viewed as the sum of these coexisting complexes. The success or failure of parasitoids to persist on their host species will ultimately shape total species richness patterns, since an extinct parasitoid is not going to generate any additional species in the future. Therefore, to understand regional richness patterns, it is useful and necessary to examine individual parasitoid complexes.

In this section I examine geographical patterns of parasitoid community richness, with the emphasis on tropical versus extra-tropical richness. Although the object is to identify latitudinal gradients, if there are any, I do not use latitude *per se* as an independent variable. Instead, I use the two climatic variables, range in annual temperature and mean low temperature in the coldest month. Both are associated with latitude (see Figs. 2.3, 2.4 on pp. 16, 17) but should be better representations of the local environmental conditions with which insects must contend than is simple latitude. There is no compelling *a priori* reason to prefer one measure over the other, so initially I will analyze both. But because the results are qualitatively very similar, I will then focus on low mean temperature to reduce redundancy.

Variation in parasitoid richness among the eight host feeding niches against climate tends to fall into two broad patterns (Fig. 3.15). Completely or partially exophytic hosts support the richest parasitoid assemblages in areas experiencing high thermal variability (Fig. 3.15*a*) and/or low winter temperatures (Fig. 3.15*b*), with richness falling as climates become more stable or with increasingly milder low temperatures. Gallers, borers and root feeders, in contrast, show either no significant climatic gradient or a slight increase in richness toward less variable/warmer climates. The pattern for leaf miners is less clear: richness tends to fall as climates become more stable (Fig. 3.15*a*) but is hump-backed for mean low temperature (Fig. 3.15*b*). But overall, exophytic hosts generally support relatively depauperate parasitoid assemblages towards the tropics, whereas better concealed hosts support similarly sized complexes everywhere or slightly richer assemblages in the tropics.

A major consideration in analyzes of geographical/climatic patterns is that the state of knowledge varies considerably in different parts of the world. The Palearctic and Nearctic regions are much better known biologically than are the other regions, and they also contain the most variable and coldest climates (see Figs. 2.3, 2.4 on pp. 16, 17). To what extent do the climatic patterns simply reflect the fact that parasitoid richness is greatest in the best studied regions? To permit comparisons both within and between better known and less well known parts of the world, I distinguished Holarctic and non-Holarctic hosts.

Patterns within the Holarctic mirror those for the entire world (Figs. 3.16*a*, 3.17*a*). This is not particularly surprising since Holarctic host species compose

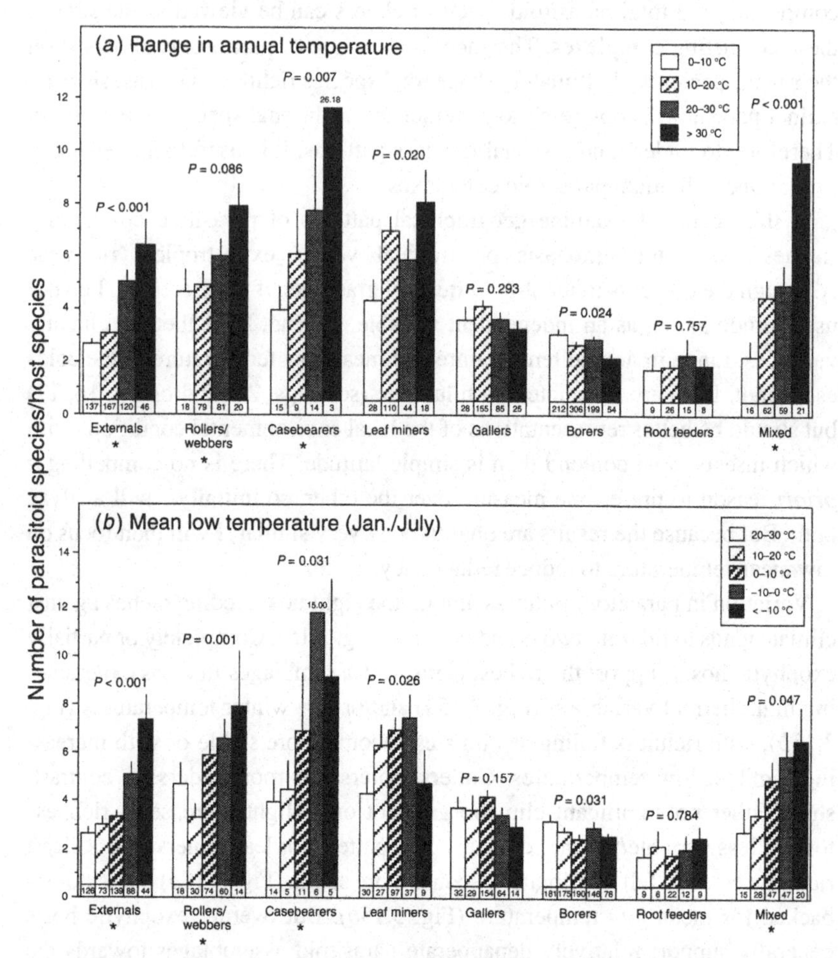

Fig. 3.15. Latitudinal gradients in mean parasitoid species richness for hosts in each feeding niche, using (*a*) range in annual temperature and (*b*) mean temperature in the coldest month as proxy variables for latitude. Within each host niche, climates are oriented with tropical systems to the left and climatically extreme temperate systems to the right. Details as in Fig. 3.1; smaller numbers above casebearers are +1 S.E.M., which extend beyond the scale of the figure.

66.7% of the global data set. But the general patterns, at least, do hold within the best studied part of the world. In the rest of the world, some patterns hold and others do not (Figs. 3.16*b*, 3.17*b*). Externally and mixed feeding hosts do tend to support more depauperate parasitoid assemblages towards the tropics, whereas endophytic hosts are similar everywhere. Thus, for these six host niches, both the Holarctic and non-Holarctic contribute to the global pattern of

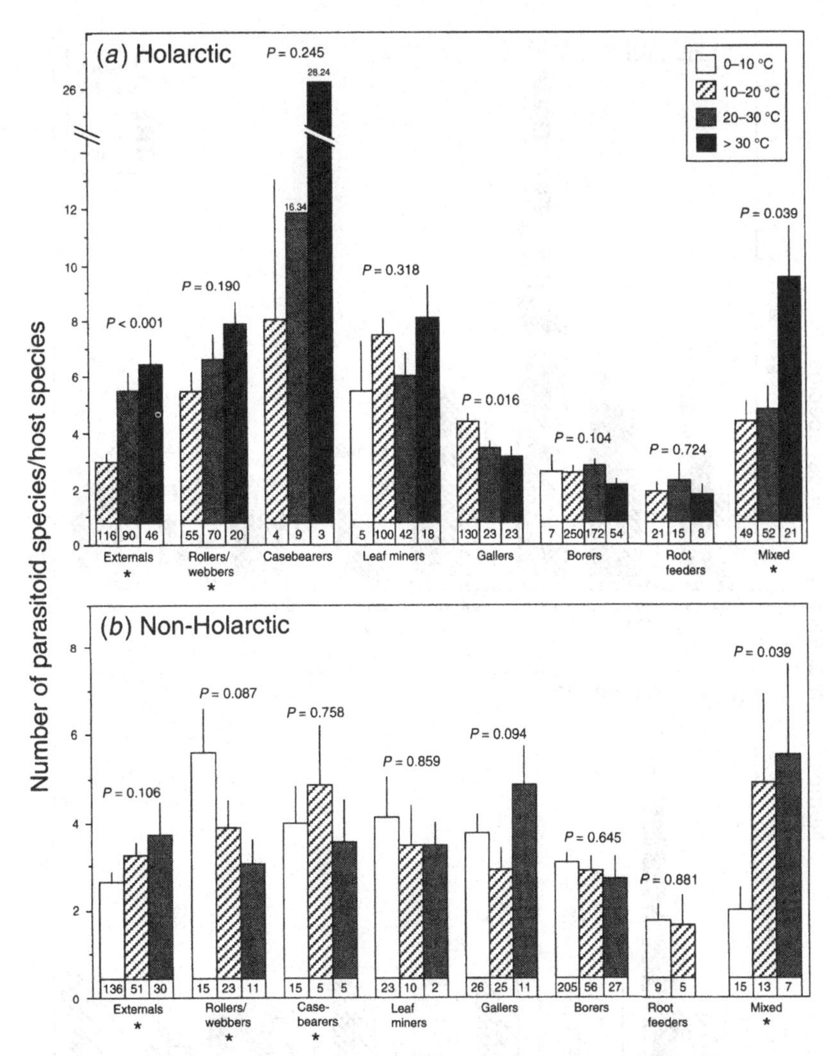

Fig. 3.16. Climatic gradients in parasitoid species richness using annual range in temperature, distinguishing (*a*) Holarctic and (*b*) non-Holarctic regions. Details as in Fig. 3.15.

falling parasitoid richness on exophytic hosts and either flat or increasing richness on endophytic hosts. On the other hand, casebearers do not vary consistently within either the Holarctic or non-Holarctic (Figs. 3.16, 3.17), and the overall pattern appears only when they are compared across regions. Finally, among leaf rollers/webbers, the patterns in the Holarctic and non-Holarctic are actually reversed for range in annual temperature (Fig. 3.16), while for mean

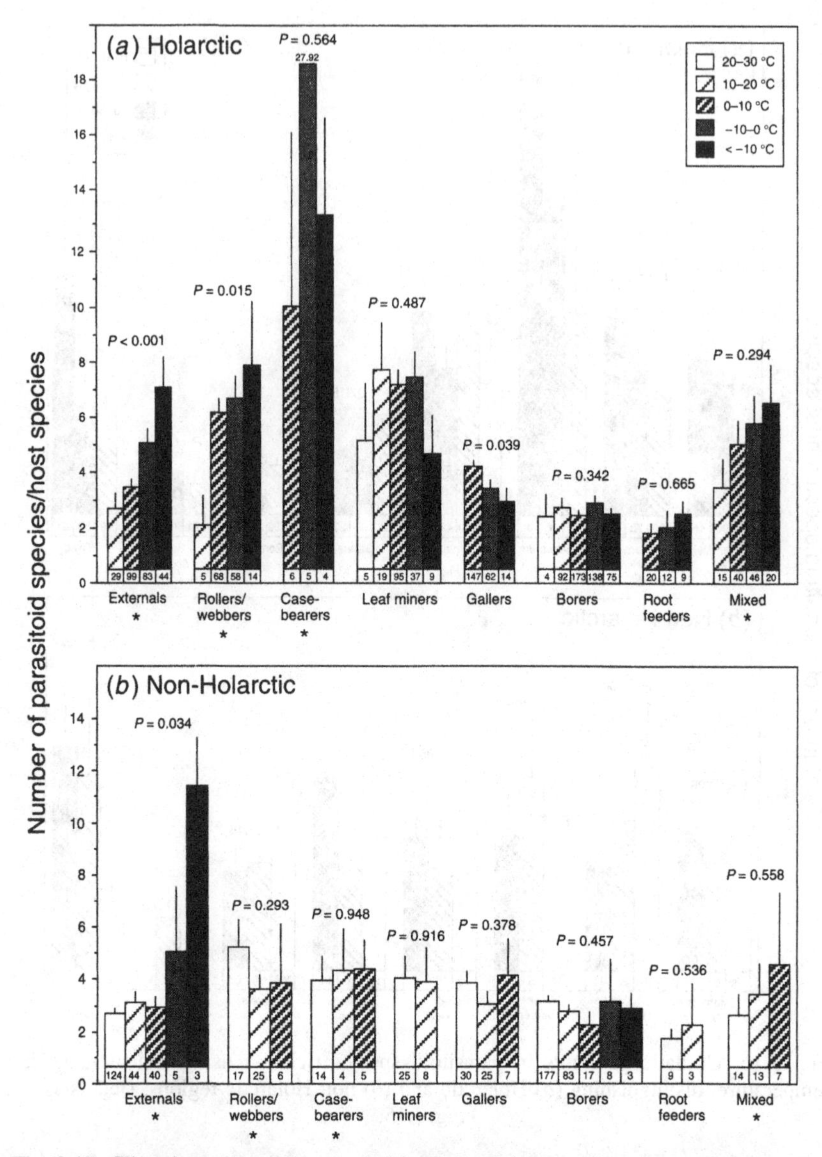

Fig. 3.17. Climatic gradients in parasitoid species richness using mean temperature in the coldest month, distinguishing (*a*) Holarctic and (*b*) non-Holarctic regions. Details as in Fig. 3.15.

low temperature, the non-Holarctic pattern is statistically flat (Fig. 3.17). For these last two host niches, it is less certain that the global patterns are not due to differences in the extent to which northern and southern parts of the world have been studied.

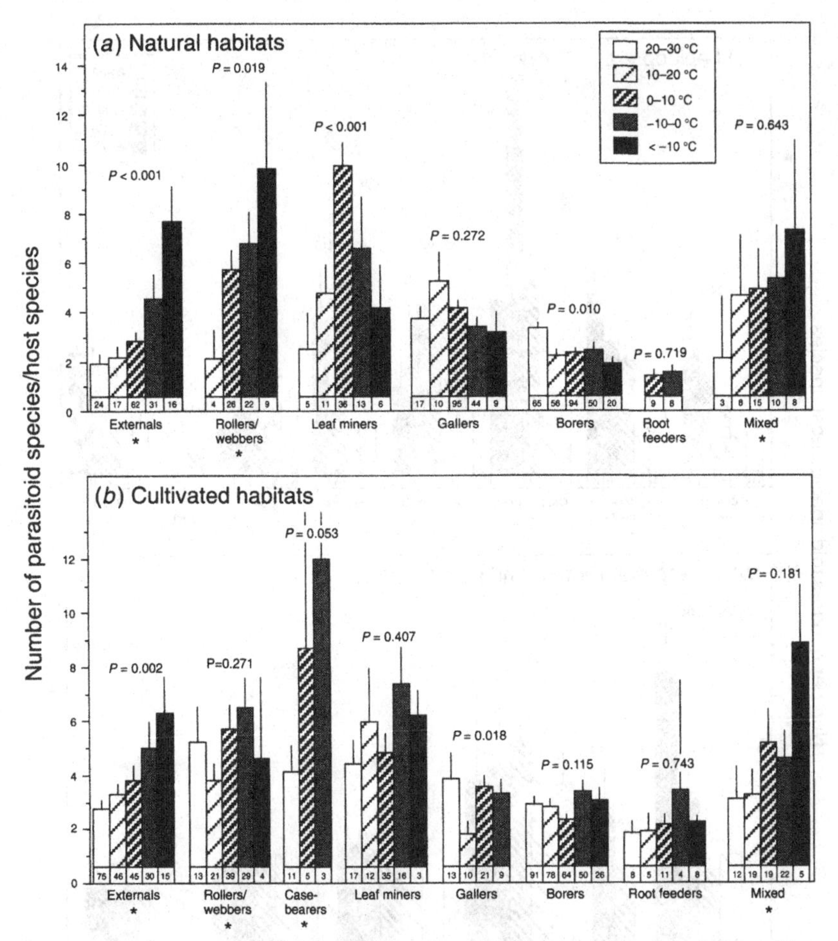

Fig. 3.18. Comparison of climatic gradients (based on mean temperature in the coldest month) in (*a*) natural and (*b*) cultivated habitats. Details as in Fig. 3.15.

There is a perception that work on tropical parasitoids has been largely restricted to agricultural systems, whereas in the temperate zone there has been much more emphasis on natural systems. Although cultivation does not generally influence parasitoid species richness greatly (Fig. 3.11), it remains possible that the climatic gradients reflect the fact that the tropical hosts that have been studied tend to be pests to a greater extent than extra-tropical hosts. However, when natural systems are examined separately, the basic form of the pattern does not change (Fig. 3.18*a*). (Because the climatic patterns are similar using either range in annual temperature or mean low temperature, only the latter is illustrated.) Parasitoid richness falls towards the tropics on partially or completely exophytic hosts, remains flat or actually increases towards the trop-

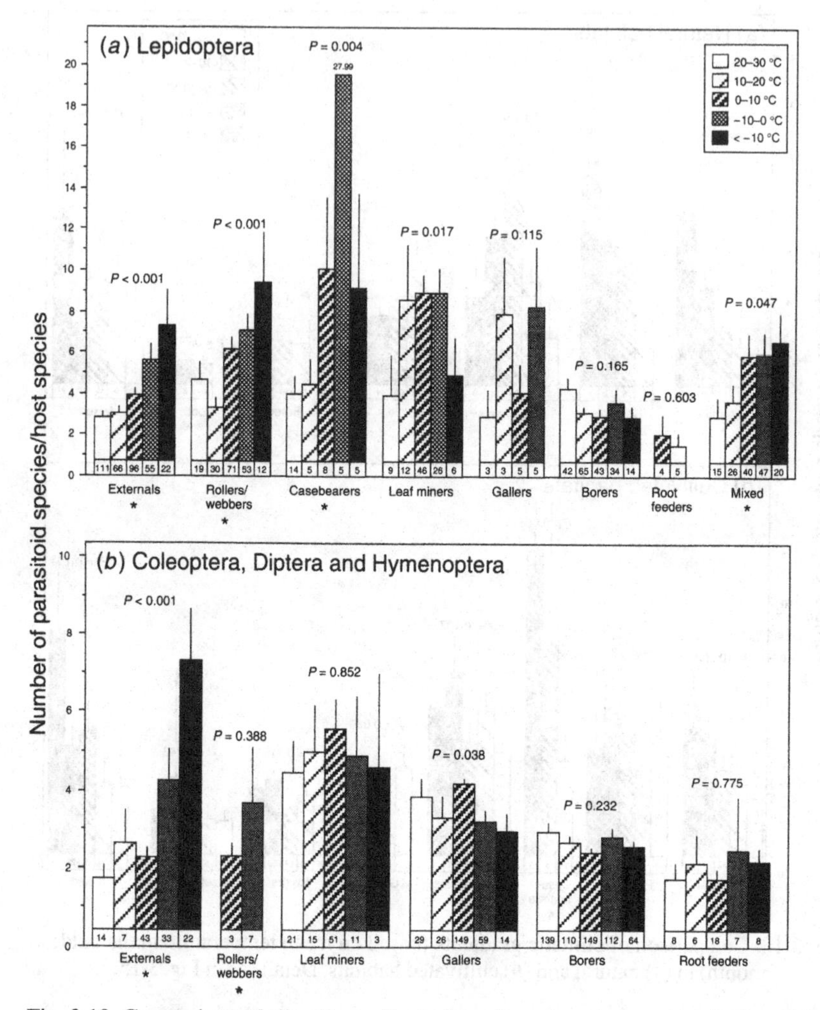

Fig. 3.19. Comparison of climatic gradients (based on mean temperature in the coldest month) for (*a*) lepidopterous hosts and (*b*) coleopterous, dipterous, and hymenopterous hosts. Details as in Fig. 3.15.

ics on gallers, borers and root feeders, and is humped for leaf miners. The pattern takes a similar form in cultivated habitats (Fig. 3.18*b*), except that there is no climatic gradient for leaf rollers/webbers or leaf miners. Therefore, for most of the host niches, the climatic gradients are reasonably robust, although more data are needed for tropical leaf rollers/webbers in natural habitats to determine what form the climatic gradient really takes.

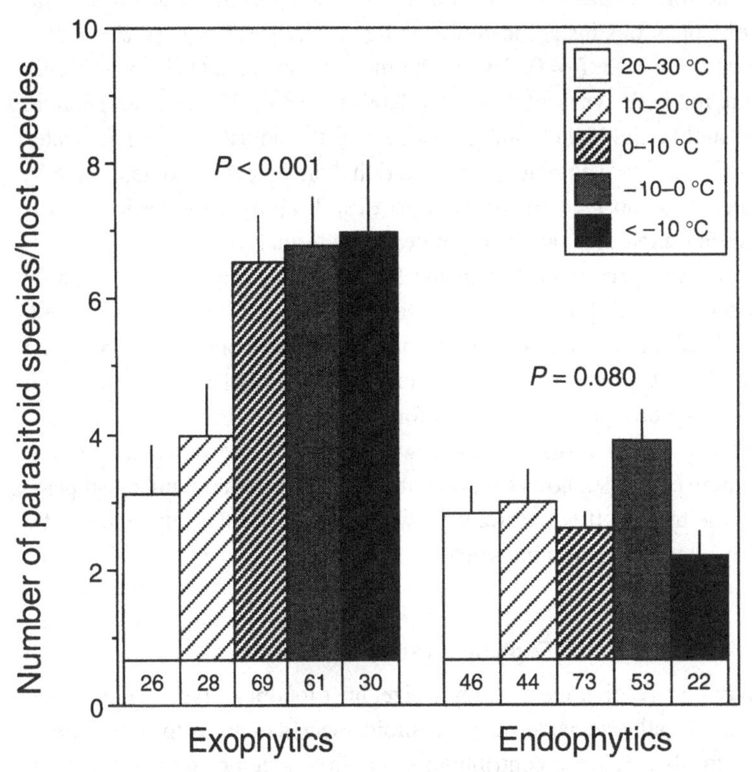

Fig. 3.20. Climatic gradients (based on mean temperature in the coldest month) for exophytic and endophytic/root feeding hosts after adjusting for sample size. Details as in Fig. 3.15.

The next question is, are the climatic gradients similar for major host taxonomic groups? Dividing the data into lepidopterous and non-lepidopterous hosts (rapidly dwindling sample sizes prevents a meaningful comparison for Coleoptera, Diptera and Hymenoptera separately) reveals no surprises; both subgroups show qualitatively similar relationships with low mean temperature (Fig. 3.19).

Finally, how do possible differences in sampling intensity in the tropics and extra-tropics influence the latitudinal gradients? Because the sample size data are less extensive, I analyzed possible sampling effects by pooling niches with similar gradients (see Figs. 3.15–3.19), resulting in 'exophytics' (i.e. external folivores, leaf rollers/webbers, casebearers and mixed feeders) and 'endophytics' (i.e. leaf miners, gallers, borers and root feeders). These were analyzed by ANCOVA with mean temperature in the coldest month as the independent variable and log-sample size as the covariate. The preliminary test for homogeneity of slopes was insignificant for both (exophytics, $F = 1.04$, $P = 0.387$; endophytics, $F = 1.53$, $P = 0.193$). In the final model, sample size was highly significant for both types of host ($P < 0.001$). For exophytics, temperature remained highly significant, and the pattern of the adjusted means indicates that tropical parasitoid complexes are less rich than temperate complexes (Fig. 3.20). There was no gradient for endophytics. Therefore, the basic patterns remain when sample sizes are incorporated into the analyzes.

The above analyses indicate that the latitudinal pattern seems reasonably robust in terms of the climatic variable used, the region of the world compared, the type of habitat that hosts occupy, major host taxonomic groupings, and sampling intensity. Two basic patterns exist, the shape of which depends on whether or not herbivores are exposed for at least part of their larval development; more exposed hosts support fewer parasitoids towards the tropics, whereas more concealed hosts support either similar or larger numbers of parasitoids in the tropics. It is uncertain to which group, if either, the endophytic but weakly concealed leaf miners belong.

3.5 Explained variance models

Host order, host feeding niche, sample size, host foodplant type, habitat type, and climate are all associated with parasitoid species richness to some extent. But what are their relative contributions, and how much of the variability in richness can be explained by these factors? I used ANOVA, linear regression and ANCOVA to examine higher-order interactions among the variables and to determine how much variance can be explained using deterministic, statistical models. I first compared the values of R^2 arising from the independent analyses of each factor considered separately, using both all available data and a subset of the data that contains information on all of the variables. I then generated multivariable models of increasing complexity to identify the groups of variables that provided the best predictors of parasitoid species richness.

All seven variables are independently associated with parasitoid species richness when all available data are used (Table 3.1). Sample size provides the best R^2, but this is based on a comparatively small number of cases. When the same host species are used to compare all of the variables, host niche explains slightly more variance than sample size or host order, and these three variables explain at least three times as much of the variance as any other factor. In the smaller data set, habitat ceases to be significantly associated with parasitoid richness.

In the multifactor models, each model includes only main factors and their interactions that are significant (i.e. non-significant interactions were

Table 3.1. *Statistical models examining the ability of seven factors to explain log parasitoid species richness, using all data and a standardized data subset restricted to hosts for which information for all variables is available* (n = *390)*

Factor	F	Probability	R^2	n
All data				
Sample size	81.27	< 0.001	0.153	452
Host niche	39.61	< 0.001	0.113	2188
Host order	50.75	< 0.001	0.065	2188
Plant type	34.89	< 0.001	0.047	2099
Mean low temperature	11.67	< 0.001	0.021	2191
Range in temperature	14.77	< 0.001	0.020	2195
Habitat type	9.11	< 0.001	0.010	1871
Standardized data				
Host niche	11.25	< 0.001	0.171	390
Host order	23.28	< 0.001	0.153	390
Sample size	68.44	< 0.001	0.150	390
Plant type	6.08	< 0.001	0.045	390
Mean low temperature	2.90	0.022	0.029	390
Range in temperature	5.70	0.001	0.042	390
Habitat type	1.40	0.248	0.007	390
Multiple factor models				
Host order × sample size[a]			0.324	390
Host niche × sample size[b]			0.322	390
Host niche × sample size × mean low temperature[c,d]			0.363	390
Host niche × sample size × range in temperature[c,e]			0.424	390
Host niche × sample size × plant[f,g]			0.406	371
Host niche × sample size × range in temperature × plant[c,f,h]			0.426	371

[a]Interaction included ($P = 0.014$).
[b]Interaction excluded ($P = 0.171$).
[c]Leaf miners and casebearers pooled.
[d]Niche × temperature interaction included ($P = 0.032$).
[e]Niche × temperature interaction included ($P = 0.002$).
[f]Root feeders excluded.
[g]Niche × plant interaction included ($P < 0.001$).
[h]Niche × plant interaction included ($P < 0.001$), other interactions excluded ($P > 0.05$).

Host feeding niche, host order, plant type, mean low temperature, and habitat type were analyzed by ANOVA, sample size (log-transformed) by linear regression, and multiple factor models by ANCOVA.

removed). Thus, each model represents the simplest of the possible models that incorporate each combination of variables. Because of limited data, certain host niches have been either excluded or pooled in particular models (see footnotes in Table 3.1).

The best two-factor models incorporate sample size with host order or host niche, and each explains about a third of the variance in richness. Incorporating either measure of temperature or plant type significantly improves the host niche model, with the model including range in annual temperature explaining the most additional variance. Adding plant type to this three-factor model significantly improves the overall model but contributes relatively little to the explained variance (explaining only an additional 2%). Habitat type does not contribute significantly to any model.

Generating a parametric model that incorporates both host niche and host order simultaneously is problematical, because not all orders contain herbivores in all of the eight feeding niches (see Fig. 3.7). The single-factor models (Table 3.1) suggest that niche is generally a better predictor than order, but after sample size is taken into account in the two-factor models, they are virtually identical. This identifies evolutionary aspects of host/parasitoid phylogeny as being about as important as ecological factors in accounting for variability in species richness. Due to a lack of the appropriate data, I am unable to pursue this further and, additionally, attempting to analyze phylogeny is beyond the scope of this work. Host order at least appears to operate independently of the ecological factors (Figs. 3.7, 3.19), so that excluding it from models does not introduce a strong directional bias. Further, in all three- and four-factor models using host order instead of host niche, those incorporating niche always gave higher R^2 values than the same models incorporating order.

In general, the models indicate that host feeding niche is the single best predictor of parasitoid species richness, but that sample size, and probably host order, are close behind. Temperature variability is also somewhat important, but only after host niche and sample size have been accounted for. Plant type, which is moderately important when considered independently, explains a statistically significant amount of the remaining variance but contributes little to actually predicting the number of parasitoid species a herbivore will support. Finally, habitat manipulation has only trivial effects on parasitoid species richness overall, probably at least partially because of the complex interactions that it has with plant type.

It is not possible to incorporate host generic diversity into the above models, because it has been quantified only in Britain. However, it does not appear to be of over-riding importance. When feeding niche and generic diversity were analyzed separately (Section 3.2), feeding niche (using mean parasitoid richness for each host genus) explained 24.0% of the variance, whereas generic diversity explained 9.4%. In the ANCOVA of feeding niche with generic diversity as the covariate, the feeding niche sum of squares was 3.3 times greater than that of diversity. The complete model explained 30.3% of the variance.

3.6 Conclusions

The various analyses of parasitoid species richness indicate that the single most consistent and powerful factor influencing how many parasitoid species a herbivore species is likely to support is the host's feeding niche. Although mean species richness patterns for each feeding niche shift relative to each other when different approaches are used and subsets of data are tested, the basic form of the dome-shaped relationship appears robust. This raises the obvious question: what mechanism(s) is (are) responsible? There are several possibilities.

In analyses of less extensive portions of these data (Hawkins & Lawton, 1987; Hawkins, 1988, 1990), I argued that the shape of the feeding niche relationship suggested that at least two interacting factors constrained parasitoid species richness, with richness increasing in the series externals—leaf rollers/webbers—leaf miners owing to decreasing host mobility, and then again falling in the series leaf miners—gallers—borers—root feeders due to increased difficulties in host location and parasitoid oviposition as hosts are better concealed and protected by plant tissues and the soil substrate. This hypothesis is basically an extension of that of Shaw & Askew (1979) and Askew & Shaw (1986) who reasoned that parasitoid richness depends on hosts being 'accessible' and 'detectable' and arises from a combination of host mobility and crypsis. However, the role of host mobility is questionable. Previous data sets did not distinguish casebearers which share characteristics with both leaf miners (encased by a relatively thin layer of plant material) and other exophytics (more-or-less mobile). That they generally support parasitoid assemblages at least as rich as leaf miners (Fig. 3.1) despite being mobile, suggests that host mobility may not be sufficient to explain the species richness patterns for the other classes of exophytics. Further doubt as to the importance of mobility comes from the leaf rollers/webbers. In several subanalyses, parasitoid richness was as great as or greater than for leaf miners (see. Figs. 3.4, 3.6, 3.7, 3.14, 3.15). Leaf rollers and their ilk are generally not as free-ranging as externally feeding folivores, but many are certainly more mobile than leaf miners.

The strongest evidence that mobility *per se* may not be behind the pattern for exophytics is based on the finding that their parasitoid complexes vary in richness dramatically with climate/latitude (Fig. 3.15). In cold/variable climate, exophytics support rich parasitoid complexes rivalling even the richest leaf-miner parasitoid complexes. I doubt that exophytics are more sedentary in the temperate zone than in the tropics, and it seems highly unlikely that exophytics support the range of parasitoid species richness that they do simply because they can move at different rates in different latitudes. To evaluate host mobility properly requires information on just how much each host moves. Unfortunately, such data do not exist, so it is not possible to test the mobility hypothesis in any detail. For now, it does appear that external feeders and leaf rollers often support fewer parasitoids than at least some classes of endophytics. It can be argued that host mobility has strong effects on parasitoids in at least some host groups (Shaw & Askew, 1979), and that host individuals may escape parasitoid attack by wriggling or dropping off the host plant (Gross, 1993), but mobility may not strongly influence parasitoid species richness in general.

Among the endophytic feeders, on the other hand, the consistent decrease in parasitoid species richness from leaf miners to root feeders (Fig. 3.1) may be largely explicable by a decrease in host susceptibility as hosts become better concealed and protected. All else being equal, the highly visible and poorly protected leaf miners should be most susceptible to parasitoid attack if host location is an important constraint on parasitoids. Consequently, we would expect miners to be highly vulnerable to parasitoid attack and to accumulate many parasitoid species. Gallers should derive some protection from attack by surrounding themselves with enlarged and often toughened gall tissue, but because they still provide specific information as to their presence (i.e. the gall), they should remain vulnerable to those parasitoids possessing specialized search behaviors, those with ovipositors long enough to penetrate gall tissues, and those that can attack the host early in its development before gall tissues fully develop. Borers should similarly benefit from being fully embedded within tough plant structures which should place many individuals beyond the reach of parasitoid ovipositors and could gain additional protection by providing less information to visually searching parasitoids. Thus, borers would be expected to support even fewer parasitoids than gallers on average.

Root feeders, representing a mix of both endophytic and exophytic feeders, occupy a habitat which should place strong restrictions on the ability of parasitoids to locate hosts relative to the other classes of herbivores. First, parasitoids must either burrow into the soil in search of potential host individuals or wait until hosts move near the surface to attack them. Second, the soil will disguise or otherwise disrupt many visual and chemical cues provided by herbivore feeding to which a parasitoid may orientate. For both of these reasons we might expect root feeders to generally have the fewest parasitoids among herbivorous hosts. Shaw & Askew (1979) have predicted that many soil insects will be shown to be entirely free of parasitoids because the latter are unable to locate hosts in the homogeneous soil environment.

Mixed exo-/endophytic feeders appear to support intermediate numbers of parasitoid species (Fig. 3.1). These hosts represent species that are concealed within either plant parts or the soil during some part of their larval developmental period, but which feed part of the time exophytically or which expose themselves as they move between endophytic feeding sites. In general, this exposure results in similarly sized parasitoid assemblages as those associated with completely exophytic herbivores. But because these herbivores represent a grab-bag of species with a wide range of relatively complex biologies, it is impossible to judge why they support the number of parasitoid species that they do.

Relative to exophytics, whose parasitoid complexes vary latitudinally, parasitoid complexes of endophytic/soil-dwelling hosts are reasonably similar in richness everywhere (Figs. 3.15–3.18). The factors that influence the susceptibility of endophytics may thus be similar no matter where they occur. I suggest that the inherent vulnerability of a leaf miner versus a galler or stem borer, arising from the supposition that leaf mines must provide less protection than a gall, plant stem or root, is sufficient to account for this. Of course, parasitoid species richness varies substantially among host species within each of the feeding niches as well, but these sources of variation are insufficient to explain the between-niche differences. My hypothesis to account for the relationship between endophytic host niches is that host susceptibility (measured as the extent that host individuals are concealed and protected by the plant parts they infest and that provide them refuges from parasitoid attack) is the primary factor constraining the number of parasitoid species that will attack them.

The susceptibility hypothesis proposes that increasing levels of concealment create stronger and stronger refuges that make deeply endophytic hosts more difficult for parasitoids to locate and better protect them against attack, even by niche-specific parasitoids that actively search the microhabitat where hosts occur. This results in either a general loss of parasitoids or, alternatively, a failure of the parasitoids to arise in the first place. This restriction of parasitoid richness can take two forms. First, moving from one mode of feeding to another can eliminate entire guilds of parasitoids. For example, exophytic sawflies support about 16 parasitoid species on average, whereas gallers support approximately four species (Price & Pschorn-Walcher, 1988). Both feeding niches support similar numbers of larval parasitoids, but stem gallers have lost almost entirely the cocoon parasitoid guild. Second, it is likely that in at least some systems increased concealment also reduces richness within each parasitoid guild. Mills (1993) found that endoparasitoids attacking early host larvae (that he called the young larval parasitoid guild) attack all types of tortricoid, but the number of species per host species declines from leaf miners to gallers.

Because exophytics are exposed, variability in susceptibility cannot be due to plant structural protection. Consequently, it is necessary to invoke at least one additional factor to account for the variability in parasitoid species richness in these groups. I am unable to examine statistically all of the possibilities, but exophytic herbivores possess a wide range of behavioral, morphological and chemical defenses against natural enemy attack (Gross, 1993) that can reduce their susceptibility to parasitoids despite them being exposed. The intrinsic defenses that have evolved in herbivores that are obviously exposed to attack by both vertebrate and invertebrate predators as well as parasitoids are probably able to account for a restriction in parasitoid species richness in a relatively straightforward way.

To this point, the susceptibility hypothesis outlined above represents little more than a just-so story to explain the observed parasitoid species richness pattern. A critical examination of the hypothesis requires additional types of data and analysis, which I will present in subsequent chapters. But before proceeding, it is worth considering an alternative hypothesis proposed by Godfray (1993) which could account for the differences between the feeding niches under a mechanism quite distinct from the susceptibility hypothesis. The 'host homogeneity' hypothesis basically proposes that the feeding niches differ systematically in the amount of homogeneity each contains. This homogeneity can be either taxonomic or ecological. Godfray hypothesized that '...the strong selection pressures exerted by feeding within a leaf lamina leads to great ecological homogeneity among leaf miners and this may explain their high parasitoid species load. In addition, relatively few taxa have evolved the leaf-mining habit and the ecological homogeneity is reinforced by taxonomic homogeneity – most leaf miners are members of large groups of related species.' He also compared leaf miners with other endophytics. 'Increasing heterogeneity is also found in the sequence of feeding niches beginning with leaf miners and moving towards more concealed hosts. Gall makers are normally members of large taxonomically homogeneous groups, yet variable gall morphology leads to ecological heterogeneity. The greatest ecological heterogeneity is probably found among borers in stems, wood and flowers, and also in root feeders.' The ecological component of Godfray's hypothesis can actually be traced back at least to Askew (1975) who observed that in contrast to cynipid-gall parasitoid communities that are strongly influenced by the diversity of gall sizes and shapes, leaf miner parasitoids are not strongly influenced, because '...there is no diversity of mine form comparable to the great variety of galls. The variation between mines involves little more than differences in mine outline and the track followed by the miner.'

This hypothesis makes several predictions that it is possible to test at least partially. First, taxonomic homogeneity is predicted to differ among the feeding niches (leaf miners are presumably more taxonomically homogeneous than borers, for example). Under this hypothesis, host taxonomic diversity should interact with feeding niche. However, in Britain, this is not the case (Section 3.2). The ANCOVA incorporating host generic diversity indicates that it does affect parasitoid richness, but its effect is *independent* of feeding niche. It is true that leaf miners from speciose genera support more parasitoids than miners from small genera, and that this relationship exists in other host feeding niches as well. But differences among the feeding niches remain even after host diversity is taken into account. Taxonomic homogeneity at the host genus level, like sample size, accounts for significant variation in species richness within feeding niches, but it is by itself insufficient to explain between-niche differences.

Diversity at the level of host order also provides little evidence in favor of the hypothesis that within-group host taxonomic diversity (measured as clade size) is of primary importance to parasitoid species richness patterns. Strong *et al.* (1984) estimated that the number of truly herbivorous species of Coleoptera and Lepidoptera are roughly equal (excluding consumers of dead or fungus-infested plant material, a food source commonly utilized by wood-boring beetles). Yet, Coleoptera support fewer parasitoid species than Lepidoptera in all feeding niches, except for root feeders (Fig. 3.7). Beetle–parasitoid complexes are generally depauperate relative to the other host orders despite the diversity of Coleoptera in all parts of the world, particularly in the tropics (Erwin, 1982). God may have an inordinate fondness for beetles, but parasitoids do not.

Hypothesized differences in ecological homogeneity among the feeding niches also make a set of testable predictions. Under Godfray's hypothesis, ecological similarity of hosts will permit extensive exchange of parasitoids among them, promoting complexes that are both rich and similar to each other in size, i.e. relatively invariable. The differences among heterogeneous hosts, on the other hand, will restrict parasitoid interchange, reproducing complexes that are less rich but which also will be more idiosyncratic, i.e. more variable. Therefore, this hypothesis predicts that mean parasitoid species richness in each niche will be negatively associated with variability in richness. Further, based on Godfray's ranking of the degree of homogeneity among endophytic feeding niches (above), leaf-miner parasitoid complexes should show inherently lower variability than those of gallers, followed by borers and root feeders.

Because of the statistical properties of all species count data, mean parasitoid species richness across the eight feeding niches would be expected to be positively associated with their variances. Therefore, to compare the inherent variability in parasitoid species richness in each feeding niche independently

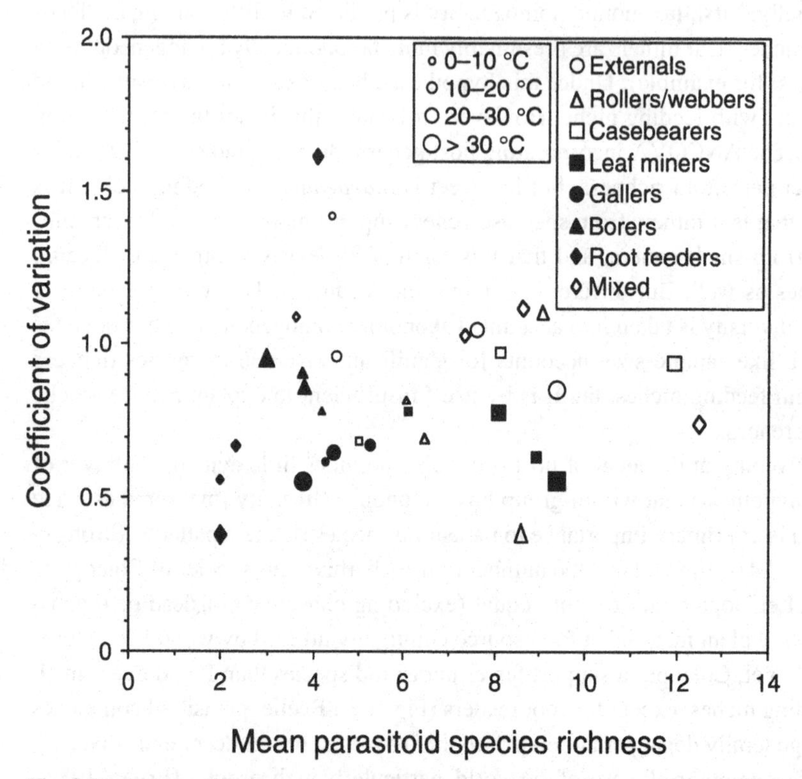

Fig. 3.21. Relationship between mean parasitoid species richness and its coefficient of variation. Hosts in different feeding niches are indicated by symbol type, and hosts in habitats experiencing different annual ranges in temperature are indicated by symbol size. See text for further explanation.

of the mean, I calculated the coefficient of variation (CV) (standard deviation/mean) for each. A problem that arises when comparing variability across feeding niches in the global data is that a major source of environmental heterogeneity has already been shown to be important. Some feeding niches show reductions in richness towards the tropics whereas others do not (Fig. 3.15). This will cause the former niches to be more variable across all environments than the latter. Although the reason for this variability has yet to be addressed, it cannot be due to the intrinsic traits of individual niches. Rather, it is related to factors operating on clusters of niches, largely depending on whether they comprise exophytic or endophytic hosts. This obvious external source of variability should be factored out to examine patterns of variability inherent to each niche. To do so, I have calculated the CVs for each niche in each range in annual temperature category. This also provides four estimates of variability within each niche that if the niche homogeneity hypothesis is correct, should on average increase along the series leaf miners—gallers—borers—root feeders.

There is little evidence of consistent patterns in the degree of variability associated with particular feeding niches (Fig. 3.21). Leaf-miner parasitoid complexes are no less variable in richness than those of gallers in any climate or root feeders in three of four cases. There is also no indication that mean diversity is negatively associated with its variability. I can find no support for the hypothesis that taxonomic or ecological homogeneity can account for niche-based parasitoid community richness patterns.

Just because I am unable to find support for Godfray's homogeneity hypothesis as a general explanation for feeding niche patterns does not imply that taxonomic or ecological heterogeneity does not influence parasitoids. They might be important in particular host groups. In an analysis of the correlates of parasitoid species richness for Nearctic and Palearctic Tortricoidea, Mills (1993) found that host feeding niche was the most important factor, but that gallers support fewer parasitoids than borers (the opposite of what the global Lepidoptera data show). He suggested that this could be explained by the relatively small number of galling species of tortricoids, because this form of ecological rarity may make it more difficult for tortricoid parasitoids to shift over to a host that has evolved a novel feeding strategy, so that heterogeneity in herbivore life-styles can influence parasitoids if it is extremely unbalanced. This seems plausible, but, even among the Tortricoidea, its general importance is still open to question. Mills found that leaf miners, another mode of feeding adopted by few tortricoids, supported more parasitoid species than hosts in any other feeding niche, despite their ecological rarity.

Although host feeding niche apparently represents the single most potent force constraining parasitoid species richness, a number of additional factors have also been found to be associated with parasitoid richness to varying degrees.

Host taxonomy almost certainly affects parasitoid species richness at all taxonomic levels. For example, Lepidoptera support the most parasitoid species, and speciose host genera support more parasitoids than small genera in Britain. Host taxonomy is a reflection of host phylogeny, and many parasitoid–host relationships are the product of physiological and behavioral constraints on parasitoid evolution and speciation mediated by host evolution. The evolutionary ecology of parasitoids has recently been reviewed (Godfray, 1993), and there is no need to address these issues in detail here. But although host evolution undoubtedly influences the nature of host–parasitoid interactions, it does not appear to be critical for detecting general parasitoid species richness patterns. Most of the major patterns arising from the analysis of ecological factors that I have examined using taxonomic subgroups appear robust. Feeding niche relationships take similar forms for all four host orders (Fig. 3.7), and latitudinal gradients do not differ for Lepidoptera versus the non-Lepidoptera (Fig. 3.19). The relationship between host taxonomic homogeneity and parasitoid species richness is also independent of feeding niche. But it is also no accident that patterns become increasingly erratic when the data are divided into finer groups and sample sizes are small. Part of this is due to the variable quality of the data, but it must also partially represent evolutionary influences acting through individual host species. Evolutionary history is undoubtedly important for understanding specific parasitoid–host systems, but when large numbers of systems are compared statistically, the effects of individual idiosyncrasies appear to be swamped by ecological processes.

In terms of its ability to account for variation in parasitoid species richness, the type of plant on which herbivores feed represents a significant, but not particularly powerful, factor (Table 3.1). Plant patterns are primarily found in natural habitats only and are detectable among exophytic niches and leaf miners, but not for gallers and borers (Fig. 3.12). Further, although plant effects are well documented for a few host groups, there are as many comparative studies that have failed to find the pattern as there are studies that have supported it (Section 3.3).

Much of the effect of plant type operates through the properties of host or plant communities rather than through intrinsic plant characteristics. The major hypothesis to account for increased parasitoid richness on hosts on trees for the groups where it has been found is that of Askew (1980), who reasoned that trees support rich, taxonomically related host-species complexes compared to herbs, and this diversity of potential hosts attracts generalist parasitoids. These generalists, being able to attack most of the host species, elevate parasitoid richness on each. This hypothesis represents the original context for the taxonomic component of the host homogeneity hypothesis presented by Godfray (1993) to account for between-niche differences, and the same logic applies. The significant positive association between parasitoid species richness and host generic diversity in Britain was predicted by this hypothesis.

The connection between plant type and parasitoid species richness, then, is that when host taxa are more diverse on trees, parasitoid richness will also be greater. This mechanism also predicts when the relationship should not exist. For host groups in which host taxonomic diversity is not associated with plant architecture, parasitoid species richness will be greater on whichever type of plant the hosts are most diverse. Most exceptions to the association of parasitoid richness with increasingly complex plant architecture have been explained using this reasoning (Askew, 1980; Hawkins, 1988; Hawkins & Gagné, 1989). If distributions of host diversity on different types of plant were the only mechanism driving plant–parasitoid patterns, then they would simply reflect the effects of evolutionary forces driving host speciation rates on different types of plant.

For herbivores, the association between plant type and species richness is at least partly due to the increased number of niches available on large, complex plants (Strong *et al.*, 1984). This is unlikely to be of any consequence to parasitoids. Most parasitoids are niche specific; leaf-miner parasitoids, for example, rarely attack exophytics, gallers or borers. Because a leaf mine is a leaf mine whether on a herb or tree, the existence of any additional structures that may support more guilds of herbivores (e.g. bark borers) is irrelevant and cannot account for an increase in species richness of leaf-miner parasitoid complexes on trees. Therefore, any increasing complexity of architecture *per se* across different types of plant is unlikely to directly contribute to the patterns.

A factor that is important involves differences in herbivore abundances on different types of plant (Askew & Shaw, 1986; Hawkins, 1988). Individual herbs are small, whereas trees are large. It is likely that, on average, herbivore populations will be largest on trees. For reasons already mentioned, abundant hosts will support more parasitoids than rare hosts. The relationship between plant type and host population size will be further enhanced when stand size is positively associated with individual plant size (i.e. if trees occur in larger stands than herbs). Even so, plant size need not be associated with stand size for the overall effect of host abundance to be manifested. Herbivores that consistently occur as large populations (whether on large individual plants or in large plant stands) provide the resource base needed for parasitoids to specialize on them and will be more often encountered by all types of parasitoid searching that habitat. Tscharntke (1992a) explained the species richness of the parasitoid complex of *Giraudiella inclusa* Fr. (Cecidomyiidae) on common reed (14 species, 12 of which are monophagous) partially in terms of the huge midge populations that develop within large foodplant monocultures. Finally, if host populations are large, they can attract parasitoids independently of the diversity of other potential host species. Mills (1993) suggested that host population abundance is more important to tortricoid parasitoid communities than is the number of alternative host species. I suspect that differences in host abundance among different types of plant will ultimately be shown to be of central importance for understanding many of the observed relationships between plant and habitat factors and parasitoid assemblage size.

The finding that parasitoid species richness increases when monocots and herbs are cultivated (Fig. 3.14) suggests that host abundance is generally important for parasitoids that normally search low plants or disturbed habitats; additional parasitoid species are able to locate hosts on herbs when resources are concentrated. But in contrast, parasitoid richness decreases when trees are cultivated. Herbivore populations on trees may naturally be large enough that further increasing them makes little difference. Instead, it appears that parasitoids are lost, due to at least three contributing factors: (a) many generalized forest parasitoids do not search in the greatly simplified plantations; (b) the reduction of nectar, pollen, and honeydew sources utilized by many adult parasitoids (Evans, 1994, and references therein); and (c) the loss of essential alternative host species required by pleurovoltine parasitoids in different times of year (Shaw, 1994). The importance of plant diversity to natural enemies has received much attention (recently reviewed by van Emden, 1981, 1990; Way & Cammell, 1981; Andow, 1991), and rearings of hosts from different plant species frequently produce different constellations of parasitoids (e.g. Krishna Ayyar, 1940; Dohanian, 1942; Gothilf, 1969; Ball & Dahlston, 1973; Nagarkatti & Ramachandran Nair, 1973; Askew & Shaw, 1974; Rathman & Watson, 1985). If plant species are removed from a habitat, the parasitoids associated with those plants may also be absent, even if their hosts are present. Although such effects should occur in all types of habitat, the impact of losses in plant diversity on parasitoid species richness appears to be most pronounced in late succession.

Finally, herbivores studied in both native and cultivated habitats often support more parasitoids than hosts studied in either habitat type alone (Fig. 3.11). Although mechanisms for macrohabitat selection by parasitoids are not well documented (van Alphen & Vet, 1986), this is what would be expected if host habitat location occurs early in the hierarchical series of behavior that parasitoids use for host location (Doutt, 1964; Vinson, 1981). Herbivores that move from natural to cultivated habitats simply exchange parasitoids keying into their natural habitat or host plant for parasitoids that are adapted to searching heavily disturbed habitats or particular crop plants.

It is becoming clear that latitudinal patterns in regional parasitoid species richness are variable, with some parasitoid groups showing no gradient whereas others become more diverse towards the tropics. A number of mechanisms invoking both abiotic and biotic processes have been proposed to explain latitudinal gradients, all of which have attempted to explain why parasitoids should not be more rich in the tropics. When species richness is examined at the individual complex level, two basic patterns emerge. Exophytic hosts support more parasitoids in the temperate zone than in the tropics, whereas endophytic hosts generally support at least as many parasitoids in the tropics, and some types of host support slightly more (Fig. 3.15). This raises the question of which mechanism(s) can simultaneously account for both patterns. The mechanisms that have been proposed all make very similar predictions for total parasitoid communities and can be distinguished only when patterns for specialist and generalist parasitoids are examined separately. The next chapter will do this, so potential explanations for latitudinal gradients based on biotic processes, such as host foodplant chemistry, herbivore diversity and abundance, parasitoid host ranges, or interactions among parasitoids and other natural enemies will be deferred until then.

On the other hand, the general pattern is by itself sufficient to suggest that abiotic processes (i.e. climate) can be discounted. Although the observed gradients are found when using climatic variables based on temperature, it is almost certain that the relationships are simply correlative rather than cause and effect. Most of the parasitoids associated with the different types of herbivore are taxonomically and biologically very similar. For example, an ichneumonid or chalcidoid attacking the pupa of an exophytic host is not very different from one attacking the pupa of an endophytic host. If temperature ranges or extremes are directly responsible for the species richness of parasitoids in different parts of the world, parasitoid communities on all types of herbivore would be expected to be similarly affected. But clearly they are not. Losses of species in the tropics are restricted to those parasitoid communities associated with particular host feeding strategies. The exophytic–endophytic dichotomy suggests that the reasons lie with the biological/ecological traits of hosts and are not some general effect on all parasitoids. Although climate obviously affects insects, just as it does all other organisms, it represents an insufficient explanation. For similar reasons, hypotheses that invoke broad historical or evolutionary differences between the tropics and extra-tropics do not appear applicable. Instead, biological or ecological processes tied to herbivore feeding biology appear critical to understanding patterns of community richness and, by inference, global diversity of parasitoid species.

In the initial analysis of correlates of parasitoid and hyperparasitoid species richness in Britain, Hawkins & Lawton (1987) could explain 22.1% of the variance in richness using a multiple regression model. The global ANCOVA models have increased this to just over 40%. This brings the amount of variation explicable by deterministic models into the range of that for herbivores on plants (Strong *et al.*, 1984). Of course, the majority of the variability remains unaccounted for. Some of this unexplained variation almost certainly arises from the wide range in the quality of the data. The data provide a hint that at least part of the unexplained variability is due to the uneven quality of the data. Host feeding niche by itself accounts for 11.1% of the variance in the complete global data. But using the data from Great Britain ($n = 176$), which is arguably the best and most consistent data, feeding niche explains 33.8% of the variance, or three times that of the global data. Similarly, host order, which explains 6.5% of the global variance, explains 17.2% of the variance in Britain, or two and a half times as much. Third, plant type, explaining only 4.7% of the global variance, accounts for 19.9% in Britain, or four times as much. If these results are representative of the remaining variables, it is possible that a significant proportion of the residual variability in global parasitoid richness could actually be explained by the factors included in these analyses, if the entire world was as well studied as Britain. Finally, the multifactor models exclude host taxonomy because some feeding niches are absent from three of the four host orders. The host's order may represent nearly as important a source of variation as feeding niche and for the most part operates independently of the latter (Fig. 3.7). If host order was incorporated into the ANCOVA models, the multiple R^2 would probably rise by at least 5% and perhaps by as much as 15%.

It is also likely that additional variability will ultimately be shown to be explicable by factors not included in my analysis. Two of the most likely variables that may be important are host geographic distribution and abundance. The geographic range of foodplants has frequently been identified as the most important determinant of herbivore species richness (Strong *et al.*, 1984), and it is extremely likely that it will similarly apply to parasitoids. Hawkins & Lawton (1987), for example, found that the geographical extent of study of British herbivores was significantly associated with parasitoid species richness. Unfortunately, that analysis did not include sample size as a covariate because of lack of data, so it was not possible to distinguish sample size effects from any biologically interesting range effects. Geographic information for hosts in my global data set is still too piece-meal to do this analysis, and I am aware of no other equivalent data sets. Furthermore, I have attempted to reduce species–area effects in these data by restricting parasitoid lists to limited geographic areas (Section 2.5), but the extent that this approach successfully removes such effects is unknown. The importance of host geographic range on parasitoid richness remains essentially unstudied and represents one area in which any data would be particularly valuable.

As I have already mentioned in the context of plant effects, host abundance is likely to have a significant effect on parasitoid richness patterns. In one of the few studies that has explicitly compared closely related host taxa, one common and one rare, Schönrogge & Altenhofer (1992) reared two larval parasitoid species over several years of sampling an extremely rare leaf-mining sawfly, whereas they recovered 11 from a common congener. Sample size differences will always confound comparisons of rare and abundant host taxa, but, even so, extremely rare hosts should support few if any strict specialist parasitoids since the probability of specialist parasitoid extinction should rise as host and parasitoid population sizes fall, even if a parasitoid could find hosts often enough to evolve specialization. Rare hosts would also be expected to support relatively few generalist parasitoids. If rare and abundant herbivores are intermingled, abundant herbivores should accumulate more parasitoid species simply because they will be encountered more frequently, unless habitats are saturated with searching female parasitoids. Even local changes in abundance over short time periods can influence parasitoid species richness. For example, the number of parasitoids reared from *Choristoneura murinana* Hübner (Tortricidae) varied between 7 and 14 species over a six-year period in a single population in Switzerland, and the number reared each year was positively correlated with budworm larval abundance (Mills & Kenis, 1991). It was assumed that the extra species reared during peak years were incidental attacks by less specialized parasitoids that had shifted over onto an abundant host resource.

That host abundance must be associated with parasitoid richness is indicated by the strong effects of sample size on perceived parasitoid richness. Sample size is at least loosely positively correlated with host abundance since on average workers will take larger samples of abundant hosts than rare ones. Unfortunately, this is confounded by how much effort was expended in sampling, so the strength of this correlation is unknown. It is not currently possible to tease apart the real effects of abundance and the 'artefact' of sampling effort. A central problem with incorporating herbivore abundances is that they are infrequently measured, and when they are, they are often not comparable across different species or types of herbivore. For example, the density of leaf miners is typically measured as the number of miners per leaf or plant, whereas for a root feeder it is usually the number per field or hectare. Until densities are available for a wide range of species and are given in a common currency, judging host abundance effects will be impossible for more than a few, restricted parasitoid communities.

Finally, and importantly, even after adding all potential taxonomic and ecological variables to analyses of parasitoid communities, it is undoubtedly the case that some proportion of the variability in species richness will always remain inexplicable. Price (1994) has discussed the myriad of forces that may act on individual host–parasitoid systems, which in concert may complicate any attempt to account for variation using general statistical models. Inherent temporal and spatial variability in size and composition is also prevalent in many parasitoid communities. These stochastic sources of variation are difficult to document in any general sense, so it will be impossible to incorporate them into analyses designed to examine the forces influencing parasitoid communities at the global level. Nevertheless, the only way forward is to attempt more detailed analyses incorporating as many variables as can be measured. Only then will we be able to evaluate the relative contributions of deterministic and stochastic processes to parasitoid community structure.

3.7 Summary

The feeding biology of hosts exerts a strong influence on the number of parasitoid species that a herbivore species supports. These effects are largely independent of a range of secondary factors, which include sampling effort, host taxonomic position and isolation, host foodplant type, habitat and latitude. For endophytic and soil-inhabiting hosts, host feeding niche may represent a measure of the susceptibility of hosts to parasitoids arising from host foodplants; increasing host concealment within, and protection afforded by, the plant parts herbivores infest provides physical refuges that make it more difficult for parasitoids to locate and attack potential host individuals. Exophytic herbivores also vary in their susceptibility to parasitoids, derived from intrinsic behavioral, morphological, and physiological defenses. This combination of extrinsic and intrinsic defenses may thus represent the most important determinant of parasitoid species richness.

Hosts on trees generally support the richest parasitoid complexes, but plant effects are confounded by the habitats in which the plants grow. Relationships between parasitoids and host plant type found in at least some host feeding niches largely disappear in cultivated habitats. There are likely trade-offs between the effects of plant diversity and herbivore abundance that may either increase or reduce parasitoid species richness in particular cultivated habitats relative to natural ones. These trade-offs appear to balance out across all types of habitat, such that hosts in both cultivated and natural habitats support similar numbers of parasitoid species on average.

Latitudinal gradients in the species richness of parasitoid assemblages largely depend on whether hosts are completely or partially exophytic, or are endophytic or in the soil. Complexes on the former are richest in the northern temperate zone and most depauperate in the tropics, whereas complexes on the latter are at least as rich in the tropics, and for some groups slightly richer. It is not possible to distinguish all of the alternative mechanisms that might account for this without more detailed knowledge of the structure of the parasitoid communities (see Chapter 4), but it is unlikely that climatic factors or historical differences between the tropics and extra-tropics are responsible.

It is possible to explain approximately 40% of the variability in parasitoid species richness in a sample of 390 parasitoid complexes with Analysis of Covariance models incorporating host feeding niche, sample size, climate, and plant type. At least some of the unexplained variance is probably due to the uneven quality of the data or to inherent stochasticity in parasitoid community size and structure, but future models may be greatly improved by incorporating additional factors, including host taxonomy, geographic distribution, and abundance.

4
Taxonomic composition and generalist versus specialist parasitoids

4.1 Introduction

The previous chapter dealt with factors influencing the total number of parasitoid species that a herbivore species may support. But not all parasitoids that compose a community are identical and interchangeable. Strict quantitative analyzes which ignore the components of parasitoid communities are likely to miss important aspects of community structure and evolution (see for example, Price & Pschorn-Walcher, 1988; Pschorn-Walcher & Altenhofer, 1989; Mills, 1993, 1994), and most of the mechanisms that have been proposed to explain large scale patterns in parasitoid species richness can only be distinguished by comparing their presumed effects on different components of parasitoid communities (Askew & Shaw, 1986; Gauld, 1986; Hawkins, Shaw & Askew, 1992). I now partition the parasitoid assemblages into component parts to examine in more detail parasitoid species richness and to evaluate some of the mechanisms that might be responsible for the patterns found in the preceding chapter.

The two criteria for dividing the parasitoids are their taxonomy and their biology. Parasitoids were distinguished taxonomically as being Hymenoptera or Diptera. Other, minor groups of parasitoids (e.g. Coleoptera) were not considered taxonomically. Within the Hymenoptera, the Ichneumonoidea, Chalcidoidea and other superfamilies were further distinguished. Parasitoids were distinguished biologically as being either idiobionts or koinobionts (*sensu* Askew & Shaw, 1986).

Parasitoid taxa are non-randomly distributed among host groups for many reasons. Parasitoid taxonomic position will be correlated with a range of other basic biological attributes which influence both host utilization patterns and parasitoid diversification (Gauld, 1988) and, consequently, almost all higher parasitoid taxa show taxonomically restricted host associations to some extent. For example, the majority of platygasterids attack Diptera, and most of those species are associated with galling Cecidomyiidae (Gauld & Bolton, 1988); most subfamilies of Braconidae are associated with individual host families or orders (Matthews, 1974; Shaw & Huddleston, 1991); and most Bombyliidae are associated with aculeate Hymenoptera (du Merle, 1975). Second, body size varies by perhaps one and a half orders of magnitude among parasitoid taxa, and some association between parasitoid and host body sizes is inevitable (see Hirose (1994) for a pioneering study of the importance of host size to parasitoid assemblages). Ichneumonids, for example, are usually much larger than chalcidoids, so although chalcidoids could develop in hosts of any size, relatively few ichneumonids can successfully develop in very small hosts. The Ichneumonoidea currently represent approximately half of all described parasitic Hymenoptera, and Chalcidoidea represent about a third (LaSalle & Gauld, 1992), so both larger and smaller parasitoid taxa are common. In the Microlepidoptera at least, endophytics are smaller than exophytics (Gaston, Reavey & Valladares, 1991), so the two parasitoid superfamilies would be expected to be differentially distributed among the host feeding niches, with chalcidoids, for example, dominating most leaf-miner parasitoid complexes (e.g. Askew, 1994). Finally, parasitoid taxonomy will be associated with the developmental syndromes of idiobiosis and koinobiosis (Haeselbarth, 1979).

The idiobiont/koinobiont dichotomy represents a potentially very informative way to view and interpret patterns in host–parasitoid interactions. At its simplest, the distinction makes strong predictions about the biological attributes of parasitoids associated with different types of host. Idiobionts, which kill or permanently paralyze their hosts during oviposition, are relatively uncommon on fully exposed hosts, because both the immobilized host and the developing parasitoid are exposed to attack by general predators and scavengers as well as to inclement weather. This forces most idiobiont groups to concentrate on concealed host species or host stages, attacking hosts feeding within plant tissues or after they have moved to a protected place to pupate. Most idiobionts are ectoparasitoids, with the main exceptions being those which attack relatively exposed, but 'armoured', host stages, such as eggs or pupae occurring on vegetation (Gauld & Bolton, 1988).

Koinobionts, on the other hand, by permitting hosts to continue to move, feed, and defend themselves are well adapted to utilize exposed hosts, although there is no physiological reason why they cannot attack concealed hosts as well. Koinobionts are most commonly endoparasitoids, which reduces the chances of the parasitoid egg or larva being dislodged or destroyed by the active host or becoming disassociated from the host during host ecdysis.

Idiobiosis and koinobiosis also make a clear prediction of the intrinsic competitive abilities of each group. Idiobionts are very frequently facultative hyperparasitoids, whereas koinobionts rarely, if ever, are (Askew & Shaw, 1986). If a host contains a developing endoparasitic koinobiont larva, subsequent attack by an idiobiont will almost invariably result in the death of the koinobiont. Koinobionts, with some exceptions, tend to attack hosts in relatively young developmental stages, depending on further host growth to supply enough nutrients for parasitoid development. Idiobionts, in contrast, must make do with the host tissues available at the time of attack, and so, except for the egg parasitoids, tend to attack later host stages. In parasitoid communities containing both koinobionts and idiobionts, koinobiont survival may depend largely on the level of subsequent idiobiont attack, unless the koinobionts are able to complete development before hosts become suitable for idiobionts. Therefore, the idiobiont/koinobiont dichotomy permits a reasonably accurate assessment of which type of parasitoid will have the edge in any competitive situation.

Idiobiosis and koinobiosis may also be associated with other basic life-history characteristics. For example, Blackburn (1991) found that idiobionts have shorter pupal periods than koinobionts, and hence shorter pre-adult life-spans. This is compensated for by generally having longer adult life-spans (Gauld & Bolton, 1988).

Finally, and importantly, classifying parasitoids by biology may represent a practical criterion of estimating otherwise largely unknown parasitoid host ranges. One justification for distinguishing idiobionts and koinobionts is that idiobionts should have broader host ranges than koinobionts (Haeselbarth, 1979; Askew & Shaw, 1986). The reasoning is simple. Koinobionts, because they have a more-or-less prolonged interaction with a physiologically active host, will be subject to host immune responses. Evolving adaptations to deal with these responses should restrict the number of hosts that koinobionts can utilize. Idiobionts, on the other hand, eliminate potential host immune responses by rapidly killing their host. This permits these parasitoids more flexibility to develop on a wider range of hosts. Therefore, idiobionts should have relatively broad host ranges (i.e. be generalists), whereas koinobionts should have relatively more narrow host ranges (i.e. be specialists). Gauld (1986), without using the idiobiont/koinobiont terminology, also argued that there is a connection between parasitoid developmental syndrome and host range as proposed by Askew & Shaw (1986).

What evidence is there to support this association between developmental syndrome and host ranges? Askew & Shaw (1986) provided a number of specific examples that support this hypothesis. More generally, they measured the host ranges of 22 idiobiont and 16 koinobiont chalcidoid species associated with arboreal leaf miners in Britain and found that the former attack on average

4.2 host families, whereas the latter attack only 1.5 host families. This initial comparison of host ranges supports their case that idiobionts should attack a broader range of host taxa, at least for this group of parasitoids.

Following Askew and Shaw's lead, several additional studies have been conducted to compare idiobiont and koinobiont host ranges. Sato (1990) also compared the chalcidoids associated with leaf-mining Lepidoptera and in Japan reared idiobionts from an average of 2.5 host genera and koinobionts from an average of 1.1 host genera. Sheehan & Hawkins (1991) expanded the test to ichneumonids by comparing Canadian rearing records for Metopiinae (a subfamily of koinobionts) and Pimplinae (an idiobiont subfamily, excluding the Polysphinctini which are ectoparasitic koinobionts attacking spiders). We found that, at the species level, pimplines had been reared from 2.3 times as many host species as metopiines and from 2.7 times as many host genera. We also found that metopiine species had been recorded from only 1 host family significantly more often than pimplines. Accounting for differential numbers of rearing records did not alter the result. Memmott & Godfray (1994) compared the host ranges of 76 species of chalcidoids attacking a wide range of dipterous, coleopterous and lepidopterous leaf miners in Costa Rica and found that koinobionts attacked an average of 1.24 host species, whereas idiobionts attacked 3.78, a highly significant difference. They also found that idiobionts attacked significantly more host families and host orders than koinobionts.

Mills (1992) has conducted the broadest based comparison of idiobiont and koinobiont host ranges, using the parasitoids of 93 species of Nearctic and Palearctic Tortricoidea. He also has taken the most detailed approach, classifying parasitoids as belonging to 11 guilds based on the host stage attacked, mode of parasitism (ecto- versus endoparasitism) and form of development (continuous versus delayed). Using this very detailed methodology, Mills found that host range depended most on the host stage attacked and parasitoid taxon and not on the idiobiont/koinobiont dichotomy *per se*. He concluded that, although the data for tortricoid parasitoids show broad support of Askew and Shaw's hypothesis, '[it] can no longer be accepted that koinobiosis and idiobiosis separate parasitoids into two classes differing in breadth of host range'. But although dividing parasitoid biologies very finely as Mills does indicate that host ranges may be constrained by multiple factors, it does not necessarily follow that the idiobiont/koinobiont dichotomy is useless. I have taken the data provided by Mills (his Table 2) and calculated mean host ranges for the hymenopterous species attacking host larvae and pupae that can be unambiguously classified as idiobionts and koinobionts. Across species in all hymenopterous groups, idiobionts attack on average 11.46 host species and 10.00 host genera, whereas koinobionts attack 6.08 host species and 5.30 host genera. Idiobiont host ranges are roughly twice as broad as koinobiont host ranges, which is close to the ratios found for the other analyses. Mills' analyses indicate that not all groups of idiobionts have broad host ranges and suggest that factors other than parasitoid developmental syndrome may partially explain them, but it does lend additional support to the hypothesis of Askew & Shaw (1986) that the dichotomy can be used to estimate average host ranges.

The only study to date that has failed to find a relationship between parasitoid developmental syndrome and host range is that of Sheehan (1991), who compared idiobiont and koinobiont Ichneumonoidea associated with folivorous Macrolepidoptera in the northeastern USA. He reported that 20.6% of 63 koinobiont species had been reared from more than one host family, whereas 40.0% of 10 idiobiont species had been. The differences are not significant (Yates corrected $\chi^2 = 0.89$, $P = 0.345$), but with only 10 idiobiont species in the data set, the test is not very powerful. It should be obvious that idiobiosis and koinobiosis cannot be used to identify any particular species as a generalist or a specialist, so when few species of either type are compared the dichotomy will be of very limited value. Mills' (1992) analyses have also shown that it is likely that not all parasitoid groups conform to the generalization. Although this is true, it does appear that for many groups and across somewhat broader comparisons idiobiosis/koinobiosis does provide a way to place groups of species in a community on the specialist–generalist continuum when detailed host range data are not available.

An additional important consideration in any discussion of parasitoid host ranges is that most species are niche specific, searching a particular microhabitat, plant part, or herbivore feeding niche (Townes, 1971; Askew & Shaw, 1986). If that niche contains few potential host species, the parasitoids will be 'specialists', but if many herbivore species occur there, there is the potential for the parasitoids to become generalists. Based on what is known of the developmental and nutritional needs of koinobionts and idiobionts (summarized by Askew & Shaw, 1986), the latter will generally be best able to take advantage of a herbivore-rich environment. Therefore, many idiobionts, although reared from few host species, are not as tied to their hosts by physiological constraints as many koinobionts appear to be. Idiobiont 'specialists' may often be so because of a lack of opportunity, and they probably represent ecological or behavioral specialists rather than physiological specialists. The entire concept of parasitoid host range is complex, and its measurement is difficult (Shaw, 1994). On the other hand, quantifying host range is essential for understanding the mechanisms driving parasitoid community organization. Although the idiobiont/koinobiont classification scheme is undoubtedly too simplistic with respect to the evolutionary, biological, and ecological determinants of parasitoid host range (see Mills, 1992) and provides at best a crude measure, alternative measures simply do not exist for most parasitoid species, particularly tropical ones.

The analysis of tortricoid parasitoids conducted by Mills (1992) also provided host range data for an important group of parasitoids for which it was unclear where they stood on the host range continuum. Askew & Shaw (1986) reported that some British Tachinidae parasitize several families or orders of hosts whereas others are apparently host specific, and they suggested that tachinids may be an exception to the general relationship between koinobiosis and host specialization found in the Hymenoptera. A similarly wide variability in host ranges is found among North American tachinids (Arnaud, 1978). On the other hand, Hagen (1964) suggested that most tachinids have broad host ranges. A lack of quantitative data on tachinid host ranges led Hawkins *et al.* (1992), in an analysis of the geographical patterns of specialist and generalist parasitoids in North America, to separate the Diptera from the Hymenoptera on the grounds that tachinids may have special host-immune avoidance and suppression mechanisms (Clausen, 1940; van Emden, 1954; Salt, 1963) that permit them to be potential generalists, even if some species have narrow host ranges.

Mills (1992) reported a mean host range of tachinids attacking tortricoids of 16.6 host species and 12.1 host genera, which is greater than idiobiont Hymenoptera as a whole, and second only to the pimplines, the hymenopterous group with the broadest average host range (27.0 host species, 23.6 host genera). Eggleton & Gaston (1992) examined the host ranges of British Tachinidae and concluded that most are relatively polyphagous and that at least some of the species that are thought to be monophagous appear so only because of limited data. Belshaw (1994) subsequently examined the known host ranges of Palearctic Tachinidae and concluded that the family as a whole is highly polyphagous. Interestingly, he also distinguished species that develop relatively slowly (i.e. are more koinobiont-like) from those that develop very rapidly (i.e. are more idiobiont-like) and found that the former have significantly narrower host ranges than those of the family as a whole, whereas the latter do not. He interpreted this as additional support for the association of parasitoid developmental syndrome with host ranges, even within the Diptera.

The currently available host range data suggest that tachinids can be broadly classified as generalists. Of course, not all species are generalists, but the same can be said of idiobiont Hymenoptera. Some hymenopterous idiobionts appear to be monophagous, and some koinobionts are polyphagous. But as long as statistical differences in idiobiont and koinobiont host ranges exist, these biological attributes can be used to examine broad patterns in parasitoid species richness and compare the predictions of hypotheses proposed to explain them. As we shall see, most of the hypotheses that have been proposed to account for latitudinal gradients in parasitoid species have predicted differences in the relative diversities of specialists and generalists in the tropics versus the extratropics. The idiobiont/koinobiont dichotomy can therefore be used as a proxy variable to evaluate which hypotheses best account for observed differences in latitudinal parasitoid community richness.

The analysis of taxonomic and biological composition of parasitoid complexes follows that used to examine total species richness, except that both data sets contain many zeros which produce highly skewed distributions. All analyzes are non-parametric, although all means reported are geometric. Finally, I examine each of the independent variables found to be associated with total richness with relation to both parasitoid taxonomy and biology, although most emphasis is placed on parasitoid biology.

4.2 Host feeding niche

Of the 12 079 parasitoid records in the complete data set, 12 067 contain at least some taxonomic information. Of the latter, 87.3% represent Hymenoptera, 12.5% represent Diptera, and 0.15% represent Coleoptera. The Ichneumonoidea compose 51.5% of the 10 405 hymenopteran records with taxonomic information, the Chalcidoidea 44.6%, and all other superfamilies 3.9%. The dipteran records are dominated by the Tachinidae (86.8%), Sarcophagidae (7.5%) and Bombyliidae (1.9%).

Parasitoid complexes on completely or partially exophytic insects are dominated by ichneumonoids, whereas leaf miners and gallers support mostly chalcidoids (Fig. 4.1*a*,*b*). Complexes associated with borers are approximately half ichneumonoids and half chalcidoids. Species from the other hymenopteran superfamilies are relatively rare in most feeding niches (Fig. 4.1*c*), being most rich on root feeders (primarily Scoliidae). As expected, ichneumonoids are most common in those feeding niches containing larger hosts, whereas chalcidoids dominate feeding niches in which hosts tend to be small.

The Diptera are most species rich on hosts that are exposed to attack by them at some stage of their larval development (Fig. 4.1*d*). Diptera lack ovipositors capable of penetrating plant tissue, so most species must depend on achieving actual physical contact with a potential host to oviposit or larviposit, or the host must be a free-ranging folivore that eats microtype eggs deposited on leaves. Consequently, leaf miners and gallers are essentially immune to attack by parasitic Diptera. Borers support a small number of tachinids. These flies typically oviposit near an entrance hole and have free-living first-instar larvae that

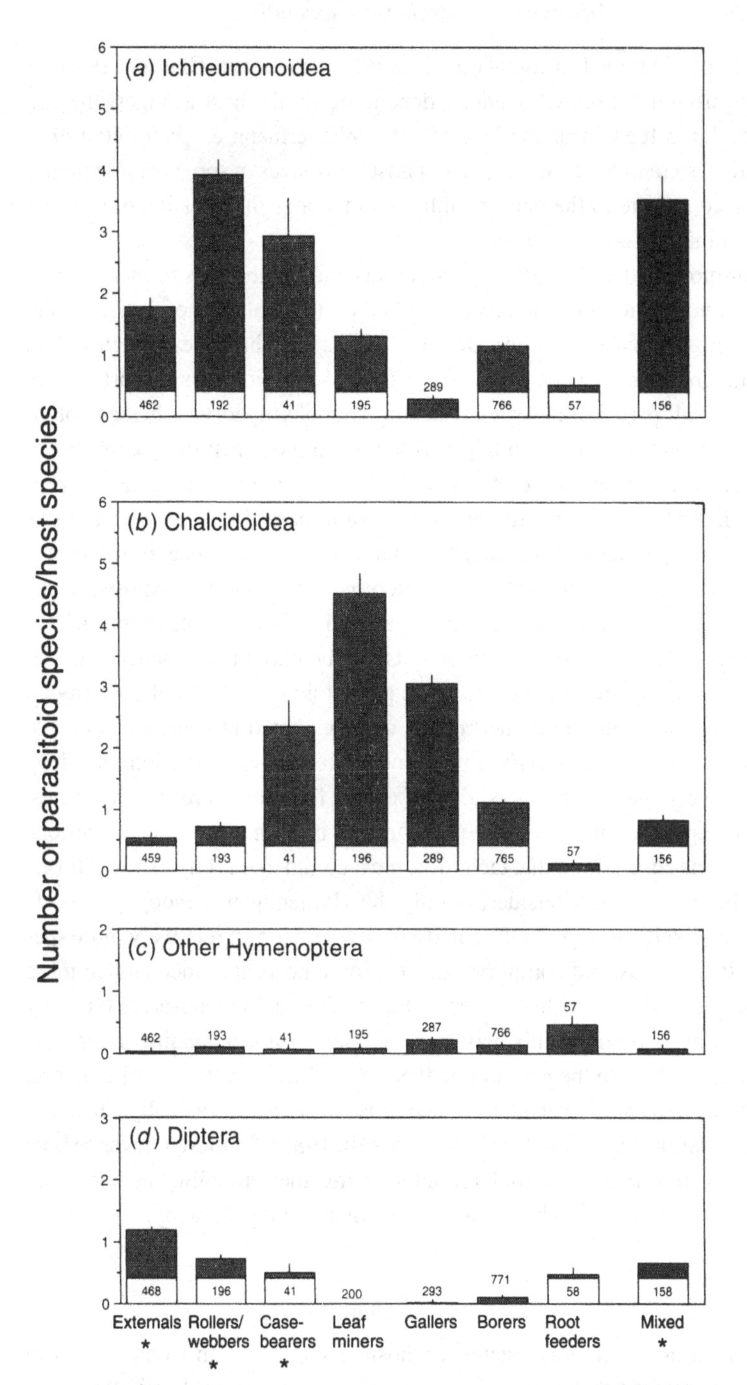

Fig. 4.1. Mean parasitoid species richness patterns across host feeding niches by parasitoid taxonomic group: (*a*) Ichneumonoidea, (*b*) Chalcidoidea, (*c*) other Hymenoptera superfamilies and (*d*) Diptera. Numbers at the base of bars are the number of host species in each niche; vertical lines are +1 S.E.M. The first seven niches are ranked in terms of decreasing mobility and increasing concealment in foodplant or soil. Asterisks indicate those niches comprising completely or partially exophytically feeding herbivores.

enter the borer's tunnel to contact the host within the plant (Clausen, 1940) or oviposit through the tunnel opening, depending on the host to ingest the egg when the larva feeds near the hole (Smith, Wiedermann & Overholt, 1993). Tachinids also tend to be large, so small host body sizes in some feeding niches probably contribute to the pattern, although this begs the question of whether or not tachinids *have* to be large.

The proportional composition of idiobionts and koinobionts in each feeding niche is concordant with that expected based on the biologies of parasitoids in both North America and Europe, the two regions for which the biologies of all parasitoid groups have been determined (Table 4.1). Idiobionts comprise more than half of all parasitoid species attacking endophytic hosts, whereas koinobionts (both Hymenoptera and Diptera) dominate the parasitoid complexes of exophytic hosts, except those of casebearers. Being within a case clearly provides sufficient protection to idiobionts to permit them to utilize casebearers as they would any other host enclosed by plant tissues. The hosts in the mixed feeding niche support relatively more idiobionts than do fully exposed hosts, but their parasitoid complexes remain dominated by koinobionts (Table 4.1).

In so far as hymenopterous koinobionts can be classified as specialists and idiobionts and Diptera as generalists, the parasitoid complexes of endophytic, casebearing and root-feeding herbivores are dominated by generalists (Table 4.1). In contrast, leaf rollers/webbers and mixed species are dominated by more relatively specialized parasitoids. External folivores, across both regions, appear to support approximately equal numbers of each. But these proportions depend on the supposition that tachinids, most common on fully exposed hosts, tend to be generalists. Considering only the Hymenoptera, endophytic hosts support relatively more generalists and exophytics support relatively more specialists. If the parasitoid complexes of exophytic hosts are indeed much more often dominated by specialists, whereas those of endophytic hosts are typically dominated by generalists, this may be important for understanding the importance of parasitoids to the population dynamics of different types of herbivore, since generalist and specialist parasitoids may have radically different dynamic relationships with their hosts (Hassell, 1986). Of more relevance here is that this may provide a tool for interpreting the variability in parasitoid species richness generated by herbivore foodplants and geography.

4.3 Foodplant/habitat

Total parasitoid richness is greatest on hosts associated with woody plants in six of seven feeding niches (see Fig. 3.10, p. 37). Hawkins *et al.* (1990) found a similar pattern for 185 herbivore species in Britain, which was due to increased

Table 4.1. *The geometric mean numbers of koinobiont and idiobiont parasitoid species per host species for hosts in eight feeding niches in North America and Europe, and the proportion of species comprising idiobionts and generalists in each niche*

	Geometric mean numbers of					
Niche	Hymenopterous koinobionts (specialists)	Dipterous koinobionts (generalists)	Idiobionts (generalists)	Proportion of idiobionts	Proportion of generalists	*n*
North America						
Externals	1.933	1.752	0.912	0.198	0.580	114
Leaf rollers	3.162	1.090	1.199	0.220	0.420	80
Casebearers	2.203	0.739	5.701	0.660	0.745	7
Leaf miners	2.177	0	4.301	0.664	0.664	58
Gallers	1.015	0.033	2.361	0.693	0.702	99
Borers	0.726	0.149	1.328	0.603	0.670	199
Root feeders	0.381	0.428	0.381	0.548	0.787	12
Mixed	2.924	0.858	1.976	0.343	0.492	70
Europe						
Externals	1.790	0.933	0.681	0.200	0.474	94
Leaf rollers	4.427	0.601	1.589	0.240	0.331	57
Casebearers	3.865	0	8.560	0.689	0.689	7
Leaf miners	2.143	0.008	4.356	0.669	0.671	91
Gallers	0.717	0.013	2.948	0.802	0.805	108
Borers	0.662	0.033	1.676	0.707	0.721	211
Root feeders	0.599	0.303	0.982	0.521	0.682	24
Mixed	2.791	0.302	1.554	0.334	0.399	40

Proportions are calculated using geometric mean numbers; *n*, number of host species.

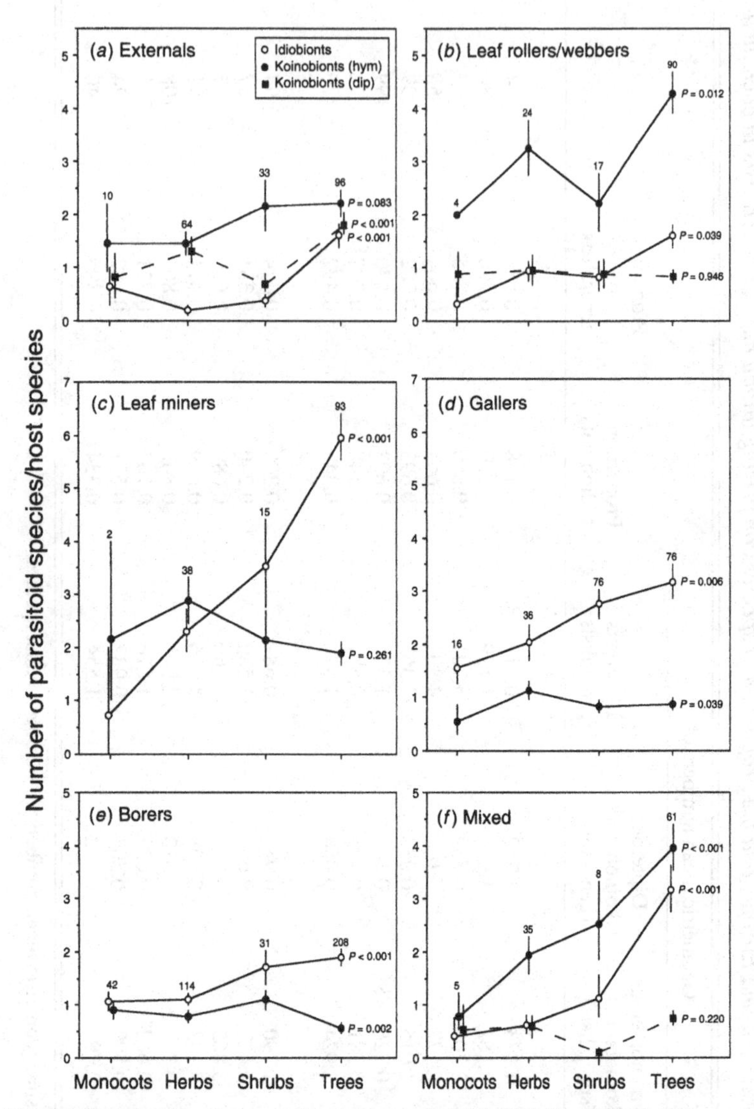

Fig. 4.2. Relationships between the mean number (±1 S.E.M.) of idiobiont, hymenopterous (hym) koinobiont, and dipterous (dip) koinobiont parasitoid species per host species and host-foodplant type in each of six feeding niches, North America and Europe combined: (*a*) externals, (*b*) leaf rollers/webbers, (*c*) leaf miners, (*d*) gallers, (*e*) borers and (*f*) mixed exo-/endophytic herbivores. Numbers above error bars are the number of hostspecies, and probabilities are from single-classification Kruskal–Wallis ANOVA.

numbers of koinobionts on exophytic hosts and idiobionts on endophytic hosts. Expansion of the data set to include 1194 species in the whole of Europe and North America results in similar, but not identical, patterns (Fig. 4.2, casebearers and root feeders excluded due to limited sample sizes). Among endophytics, idiobionts increase in species richness from monocots to trees, whereas hymenopterous koinobionts either do not change or actually decrease with increasingly complex plant architecture. For exophytics (externals and leaf rollers/webbers), on the other hand, hymenopterous koinobionts tend to be richest on trees, although not quite significantly so for externals. But, unlike the result found in the analysis of British herbivores, idiobionts are richest on trees as well, so the relatively rich parasitoid complexes of exophytics on trees comprise more species of both types of parasitoid. This pattern is even more apparent on mixed exo-/endophytics, with both koinobionts and idiobionts showing strong plant influences (Fig. 4.2). Therefore, across all feeding niches, idiobionts are richest on trees. On completely or partially exophytic hosts, koinobionts tend to be richest on trees, whereas on endophytic hosts there are no more or even fewer species.

Patterns for dipterous koinobionts appear weak at best. On externals, where Diptera richness is greatest overall, richness is highest on trees, but the progression from monocots to trees is erratic (Fig. 4.2). Among leaf rollers/webbers and mixed hosts, Diptera are equally rich on all plant types.

The overall pattern across all groups of parasitoids is that herbivores on trees support more species of generalists (including Diptera) than herbivores on herbs in all feeding niches. Specialist richness is also greater on trees for exophytic hosts, but is not associated with plant architecture on endophytic hosts. These interactions among different types of parasitoid with feeding niche and plant type indicate that the general dominance of exophytic hosts by koinobionts and endophytic hosts by idiobionts requires some qualification. Koinobionts dominate the parasitoid complexes of exophytic hosts on all plant types, but the dominance of endophytic hosts by idiobionts is strongest on woody plants and may be either absent or the reverse may be true on grasses and herbs (see also Askew (1994) for examples among the parasitoid complexes of herb- and tree-dwelling leaf miners).

Cultivation largely decouples the relationship between plant architecture and species richness for most feeding niches (see Fig. 3.12, p. 40), presumably by disrupting ecological processes generated during succession. We would expect patterns for the components of parasitoid communities to be similarly affected. The question is, are some types of parasitoid more affected by cultivation than others? There is indirect evidence that the parasitoids attacking herbivores in cultivated habitats are not the same species that attack them in

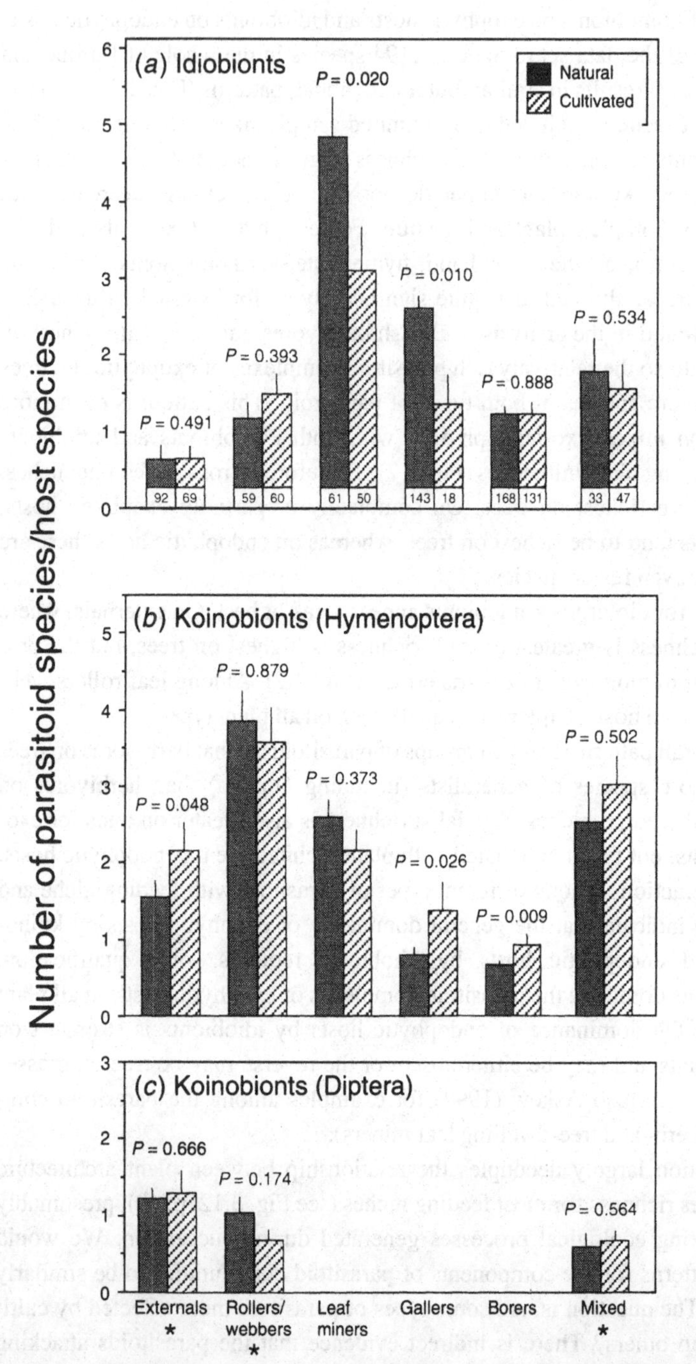

Fig. 4.3. Relationships between mean parasitoid species richness and habitat type for (*a*) idiobionts, (*b*) hymenopterous koinobionts, and (*c*) dipterous koinobionts. Details as in Fig. 4.1.

natural habitats (see Fig. 3.11, p. 38). Do the parasitoid complexes associated with hosts in cultivated habitats represent largely generalists which enter the habitat opportunistically, or do relatively specialized species follow hosts into habitats where they may not 'naturally' occur?

Comparing the richness of idiobionts and koinobionts in natural versus cultivated habitats (Fig. 4.3) reveals no significant differences in most niches. In the niches where differences are found, fewer idiobionts are associated with hosts in cultivated habitats than in natural habitats (Fig. 4.3*a*), but the number of hymenopterous koinobionts is higher in the former (Fig. 4.3*b*). Numbers of dipterous koinobionts are similar in both habitat types. Across all feeding niches, the number of idiobionts is higher in natural habitats than in cultivated habitats (mean = 1.75 and 1.40, respectively, $P = 0.009$), whereas the converse is the case for hymenopterous koinobionts (1.30 and 1.89 species, respectively, $P < 0.001$). There is no difference for Diptera ($P = 0.297$). It appears that cultivation slightly enhances the species richness of 'specialists' and reduces the number of 'generalists'!

Both plant architecture and cultivation influence the specialist and generalist parasitoids differentially, although the latter effect is weak. What about the interaction of these factors? Plant effects in natural habitats are very similar to those found across all habitats (Fig. 4.4). Idiobionts and hymenopterous koinobionts both tend to increase with increasingly complex plant architecture on exophytic hosts, with trees supporting the most species. On endophytic hosts, idiobiont richness increases consistently with increasingly complex plant architecture, whereas koinobiont richness is similar on all plant types. Diptera either increase with plant architecture or do not change.

In cultivated habitats plant effects are much weakened (Fig. 4.5), as the analysis of total species richness has already indicated. On endophytic hosts, idiobionts no longer dominate the parasitoid communities so strongly, particularly among gallers and borers. Even so, there is still a tendency for idiobionts to be relatively more important on trees than on other types of plant (Fig. 4.5). On exophytics, there are no simple patterns. Neither external feeders nor leaf rollers/webbers show convincing relationships with plant architecture, but the pattern for mixed species appears similar to that found in natural habitats.

Summing up a complicated set of patterns, the most relevant results are: (a) hymenopterous generalist (idiobiont) and specialist (koinobiont) parasitoids respond differentially to plant architecture, generalists become increasingly more important with increasingly complex plant architecture on all types of host, but specialists increase in richness only on exophytic hosts; (b) cultivated habitats tend to support relatively more specialists; and (c) plant effects are strongest in natural habitats, although the basic pattern of relatively greater

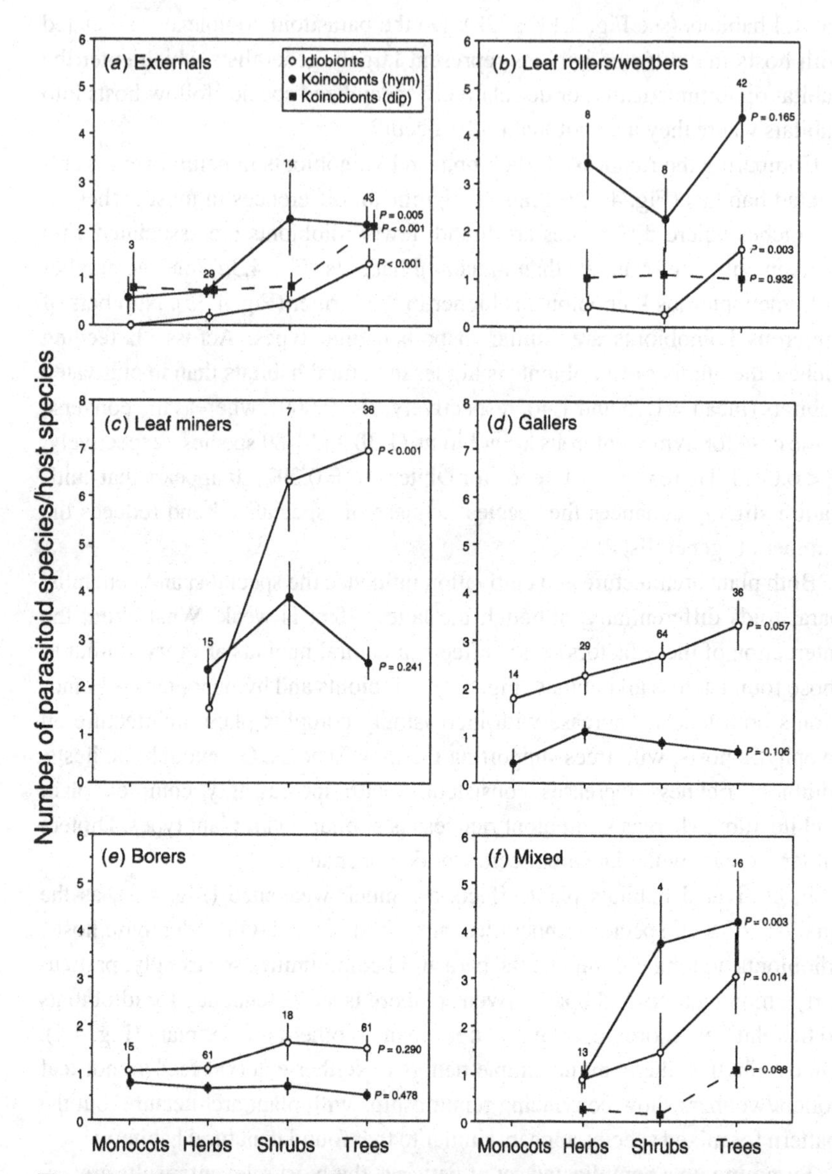

Fig. 4.4. Relationships between mean numbers of idiobiont, hymenopterous koinobiont, and dipterous koinobiont parasitoid species per host species and host-foodplant type in each of six feeding niches for hosts studied in natural habitats: (*a*) externals, (*b*) leaf rollers/webbers, (*c*) leaf miners, (*d*) gallers, (*e*) borers, and (*f*) mixed exo-/endophytic herbivores. Details as in Fig. 4.2.

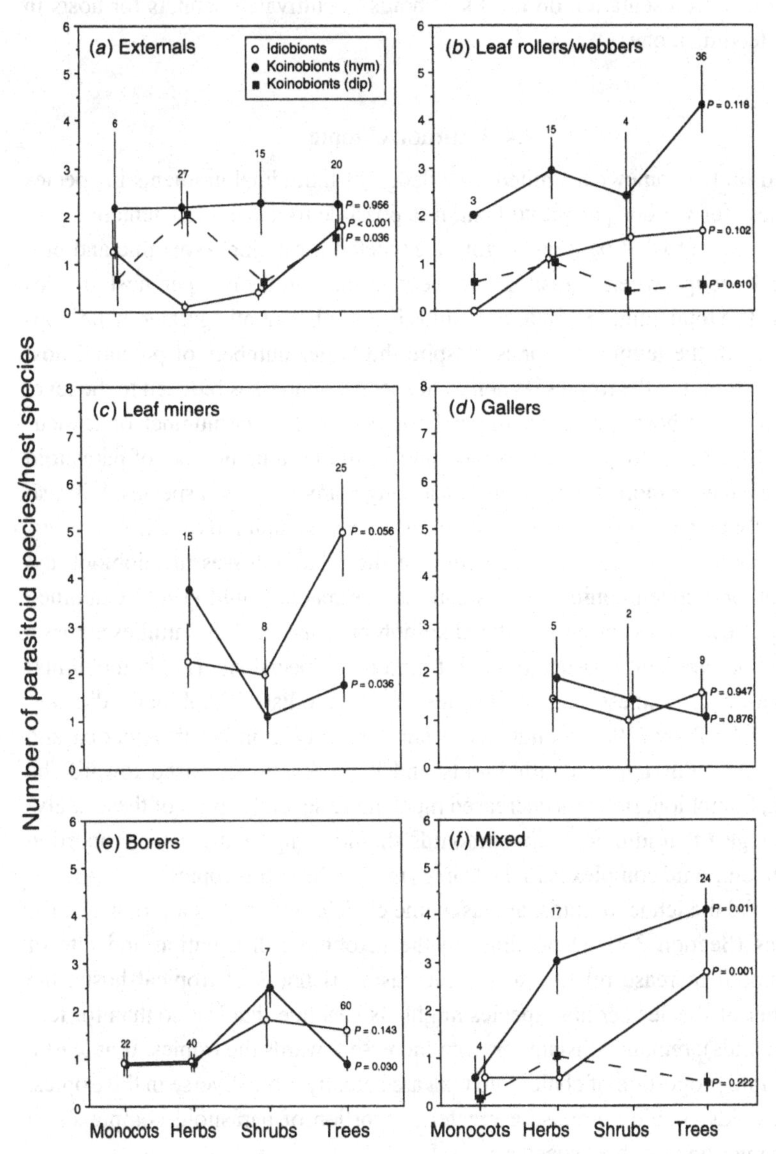

Fig. 4.5. Relationships between the mean numbers of idiobiont, hymenopterous koinobiont, and dipterous koinobiont parasitoid species per host species and host-foodplant type in each of six feeding niches for hosts studied in cultivated habitats: (*a*) externals, (*b*) leaf rollers/webbers, (*c*) leaf miners, (*d*) gallers, (*e*) borers, and (*f*) mixed exo-/endophytic herbivores. Details as in Fig. 4.2.

generalist representation on trees still holds in cultivated habitats for hosts in most feeding niches.

4.4 Latitude/climate

Based on the somewhat limited knowledge of latitudinal gradients in species richness for various parasitoid taxa, it is possible to make a few tentative predictions about what patterns might be expected for major taxonomic and biological groups when parasitoid richness is measured on a per host species basis. The total number of ichneumonids is not substantially greater in the tropics than in the temperate zones, despite the larger numbers of potential host species found in the tropics. When total species richness is reduced to the level of individual host species, it might be expected that the number of ichneumonid species is lower in the tropics since a similar total number of parasitoid species may be more widely scattered among many more host species. It is also likely that many more tropical hosts support no ichneumonids at all.

A slightly more detailed prediction of the relative losses in idiobiont and koinobiont ichneumonid richness can also be made. Gauld (1986) examined the species richness of idiobiont and koinobiont tribes and subfamilies across a latitudinal gradient in Australia and found that 'specialists' (i.e. koinobionts) decreased in richness into the tropics but 'generalists' (idiobionts) did not. Askew & Shaw (1986) conducted a similar analysis in North America and found that although both idiobionts and koinobionts decreased toward the south, koinobiont richness decreased much more severely. Both of these analyses suggest that idiobiont ichneumonids should compose a greater proportion of ichneumonid complexes in the tropics than in the extra-tropics.

Unlike the ichneumonids, at least some chalcidoid families are richer in the tropics (Section 3.4). Depending on the taxonomic distribution and rate of chalcidoid increase relative to the increasing richness of tropical hosts, the number of species per host species might also decline (but less so than for ichneumonids), remain constant, or even increase towards the tropics. If at least a moderate proportion of chalcidoid taxa are actually more diverse in the tropics, chalcidoids should compose a greater proportion of parasitoid complexes in the tropics on a per host species basis.

Latitudinal gradients of total species richness of parasitic Diptera are poorly documented, so it is more difficult to judge *a priori* what the pattern of species richness per host species might look like. Janzen (1981) predicted that tachinid species richness will be lower in the tropics, presumably because relatively more tropical herbivores roll leaves or are endophytic, making them inaccessible to dipteran attack. Askew & Shaw (1986) reasoned that Janzen might be right, but that the drop in tropical species richness of Diptera should be less than that of koinobiont ichneumonid groups, because the former's ability to expand their host ranges in the tropics (i.e. convert from being 'potential' to 'actual' generalists) should permit them to maintain their species richness to a greater extent. I am not aware of any examinations of the tropical versus extra-tropical total species richness patterns of Diptera, but Hawkins *et al.* (1992) tested Askew and Shaw's (1986) prediction using North America parasitoid complexes and found that whereas Hymenoptera were less species rich on exophytic hosts towards the south, the numbers of Diptera species per host species were similar everywhere and composed an increasing proportion of the koinobiont component of parasitoid complexes towards the south.

How do the data fit these simple predictions for each parasitoid group? Ichneumonoid species richness against mean low temperature shows significant decreases in six of the eight feeding niches (Fig. 4.6*a*). Even in the two niches for which differences are not significant, the trend is similar. There is a clear indication that the ichneumonid component of parasitoid assemblages falls towards the tropics on all types of host, irrespective of the latitudinal gradient for the entire assemblage.

The pattern for chalcidoids is not so clear (Fig. 4.6*b*). For externals, leaf rollers/webbers, root feeders, and mixed species there is no statistical gradient in richness; in casebearers there is a drop in tropical richness; in leaf miners there is a statistical relationship with temperature, but it is not a simple gradient; and in gallers and borers richness increases marginally towards the tropics. Overall, chalcidoids tend to be as rich on tropical hosts as extra-tropical hosts, and in some cases they are probably even richer.

As is typical of the analyses of Diptera, the patterns for flies are messy (Fig. 4.6*c*). Diptera species richness does drop towards the tropics on external and leaf-rolling herbivores but does not in the other niches. Since most species are concentrated on the first two types of host, Diptera probably are less rich in the tropics overall, but this loss of richness is not as strong as that found for the ichneumonoids.

The overall taxonomic composition of parasitoid assemblages is what would be expected based on known or presumed geographical patterns of parasitoid and herbivore diversity. It is safe to assume that herbivore diversity is much greater in the tropics. Ichneumonoids (including both Ichneumonidae, for which the data are good, and Braconidae, for which latitudinal data are very limited (Wharton, 1993)) are as rich or slightly less rich in the tropics, many chalcidoids are richer in the tropics than in the extra-tropics, and Diptera richness may fall towards the tropics, but less so than in ichneumonoids. If broad patterns of parasitoid host ranges do not differ too substantially in different lat-

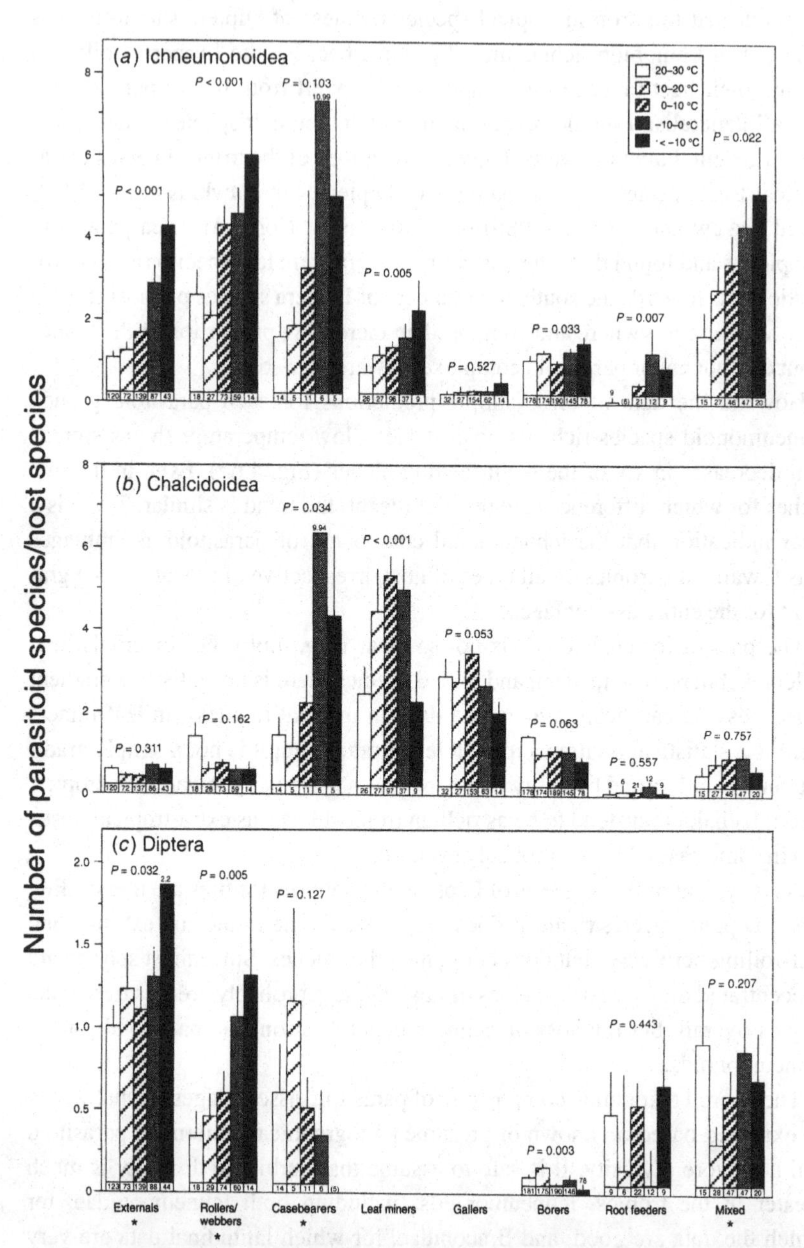

Fig. 4.6. Mean parasitoid species richness by mean temperature in the coldest month for major parasitoid groups: (*a*) Ichneumonoidea, (*b*) Chalcidoidea, and (*c*) Diptera. Within each host feeding niche, tropical habitats are represented on the left and climatically extreme temperate habitats are to the right. Note the change of scale in (*c*). Other details as in Fig. 4.1.

itudes, we would expect that in terms of individual host species, the number of ichneumonoids per host species should fall towards the tropics, chalcidoids should for the most part maintain their diversity, and Diptera richness should fall slightly. These predictions are met when comparing the mean numbers of each parasitoid taxon per host species across all feeding niches. From the lowest mean temperature zone (northern temperate) to the highest (tropics), the geometric mean numbers of ichneumonoids are: 2.443, 1.940, 1.289, 1.270, 1.015; of chalcidoids are 0.861, 1.276, 1.562, 1.059, 1.210; and of Diptera are 0.488, 0.435, 0.288, 0.358, 0.360. It appears that general patterns of regional species richness are reasonably well reflected in localized species richness patterns among individual parasitoid complexes.

Because each parasitoid taxon has a different latitudinal gradient, we would expect that each achieves a different proportional representation in parasitoid complexes across the climatic zones. Indeed, the mean proportion of the total parasitoid complexes comprising Ichneumonoidea is low in the tropics in all host-feeding niches (Fig. 4.7*a*), chalcidoids either maintain or increase their relative richness in the tropics (Fig. 4.7*b*), and Diptera show a mixed pattern (Fig. 4.7*c*). It is interesting that in the parasitoid complexes of externally feeding herbivores, the niche supporting the most species of Diptera overall, they are proportionately better represented in the tropics than in the temperate zone.

The analysis of koinobiont and idiobiont representation across climatic gradients requires two approaches. The classification of all parasitoid groups by developmental syndrome was possible only for North American and European complexes. At the global level, classifications were done for only the ichneumonoids and dipterans, because the biologies of many of the tropical chalcidoids were unknown. So the initial and most complete examination of latitudinal patterns must exclude tropical parasitoids. But because the species richness gradients appear relatively consistent across all climatic zones, and not simply tropical versus non-tropical dichotomies (but see Fig. 3.20, p. 51), comparisons within geographic regions that include cold temperate to subtropical habitats should provide insights into possible mechanisms responsible for the general global patterns.

If the comparisons of the latitudinal species richness patterns of idiobiont and koinobiont ichneumonoids (Askew & Shaw, 1986; Gauld, 1986) are representative of all parasitoid groups, we would expect koinobiont species richness to fall towards the tropics, whereas idiobiont richness should either remain flat or fall less strongly. This would be expected to occur in most or all host-feeding niches, since it is characteristics of the parasitoids themselves (i.e. their host ranges) rather than characteristics of hosts that are presumably responsible for the pattern. But within North America and Europe, this is not the case (Fig. 4.8). Among endophytic hosts (including in this case mixed exo-

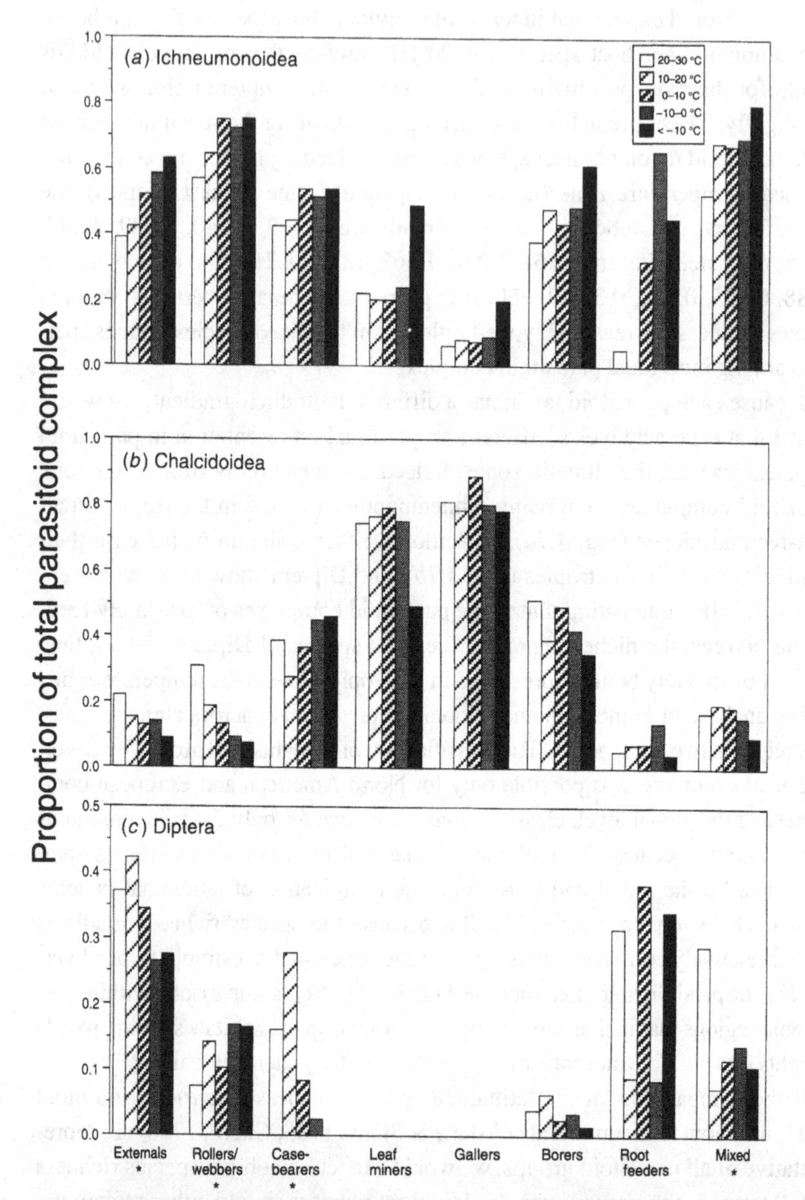

Fig. 4.7. Proportional representation of major parasitoid groups across mean temperatures in the coldest month: (*a*) Ichneumonoidea, (*b*) Chalcidoidea, and (*c*) Diptera. Note the change of scale in (*c*). Other details as in Fig. 4.1.

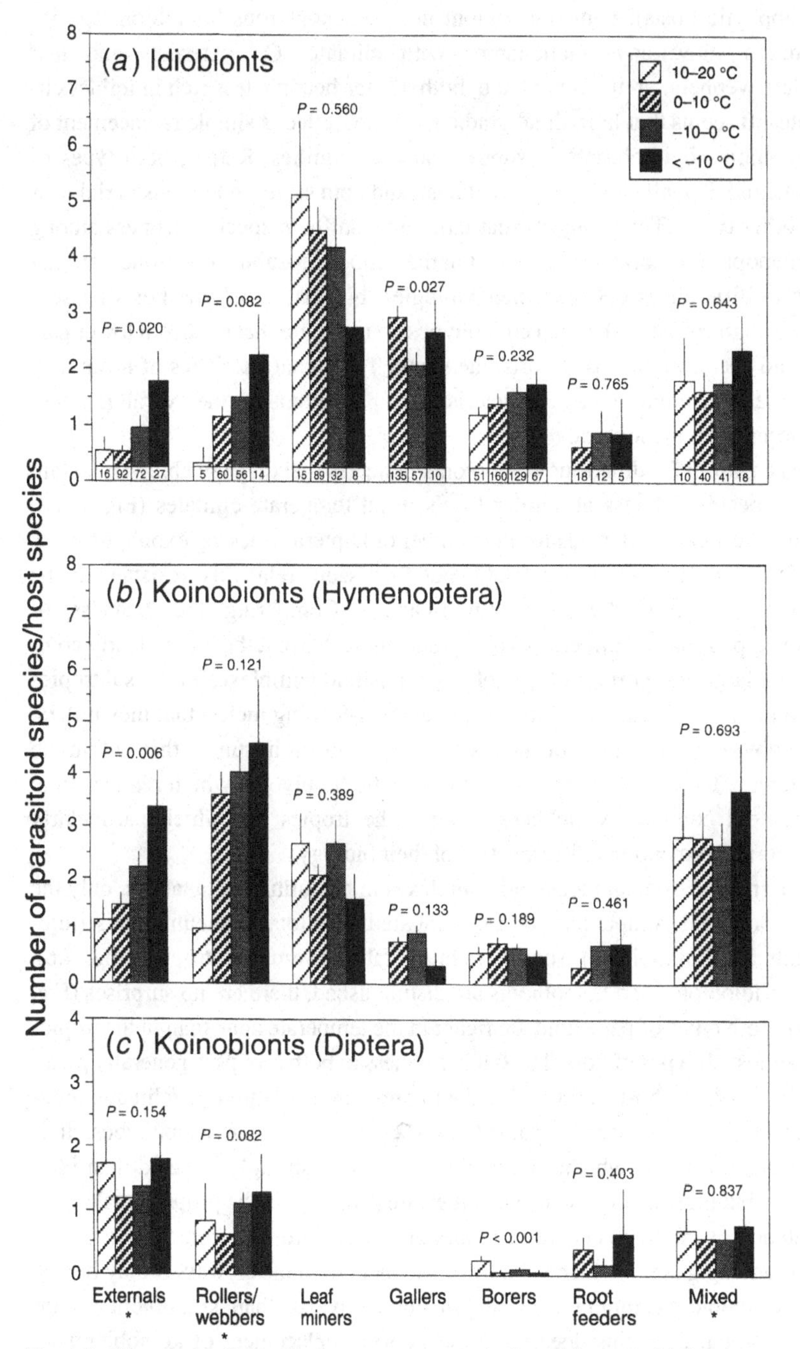

Fig. 4.8. Climatic gradients (by mean temperature in the coldest month) for components of Holarctic parasitoid complexes: (*a*) idiobionts, (*b*) hymenopterous koinobionts, and (*c*) dipterous koinobionts. Details as in Fig. 4.6.

/endophytic hosts), neither idiobiont nor hymenopterous koinobiont species richness show any relationships with climate. On externals and leaf rollers/webbers, on the other hand, both groups become less rich in milder climates. It seems that latitudinal gradients do not reflect a simple replacement of koinobionts byidiobionts in subtropical communities. Rather, both types of parasitoid do well on some types of host, and both suffer reductions in richness on other types. This suggests that latitudinal shifts in species richness among hymenopterous parasitoids, at least in the temperate–subtropical zones, are not due to differences between their biologies, but that exophytic hosts possess some characteristic(s) which endophytics do not share that results in lower parasitoid species richness towards the south. Thus, characteristics of hosts may be at least as important as characteristics of parasitoids for the overall richness patterns of Hymenoptera.

The pattern for dipterous koinobionts indicates that they are able to maintain their species richness at similar levels in all temperate climates (Fig. 4.8*c*). There are weak tendencies for the number of Diptera attacking exophytic hosts to fall slightly towards the south, but flies seem relatively resistant to the factors restricting hymenopteran richness. Comparing the richness of hymenopterous and dipterous koinobionts indicates that the latter clearly comprise a larger proportion of koinobiont parasitoid complexes in the subtropics than in the northern temperate zone in all host-feeding niches that they utilize. The obvious implication of the above patterns taken in sum is that within the Holarctic, there is something about exophytic herbivores which reduces total parasitoid community richness towards the tropics but which particularly affects all Hymenoptera irrespective of their biology.

To include tropical parasitoid complexes in the latitudinal analysis, only the ichneumonoid component can be compared. For total ichneumonoids (idiobionts and koinobionts), species richness falls towards the tropics (Fig. 4.6). When idiobionts and koinobionts are distinguished, there are no surprises (Fig. 4.9). Both types of parasitoid are richer in the temperate zone than in the tropics on almost all types of host. The fall in richness in both groups is generally greatest in exophytic hosts, reinforcing the impression that latitudinal/climatic influences are stronger on the parasitoids of exophytics than endophytics. It is important to know whether koinobionts are more strongly affected than idiobionts, but there is no evidence that they are (Fig. 4.9*c*). The proportional representation of idiobionts is not generally greater towards the tropics, except perhaps among gallers, which support few ichneumonoids anyway. It may be that many idiobiont subfamilies/tribes are more diverse than koinobiont groups in the tropics, but this does not translate to a replacement of koinobionts by idiobionts in individual parasitoid communities there. The extremely

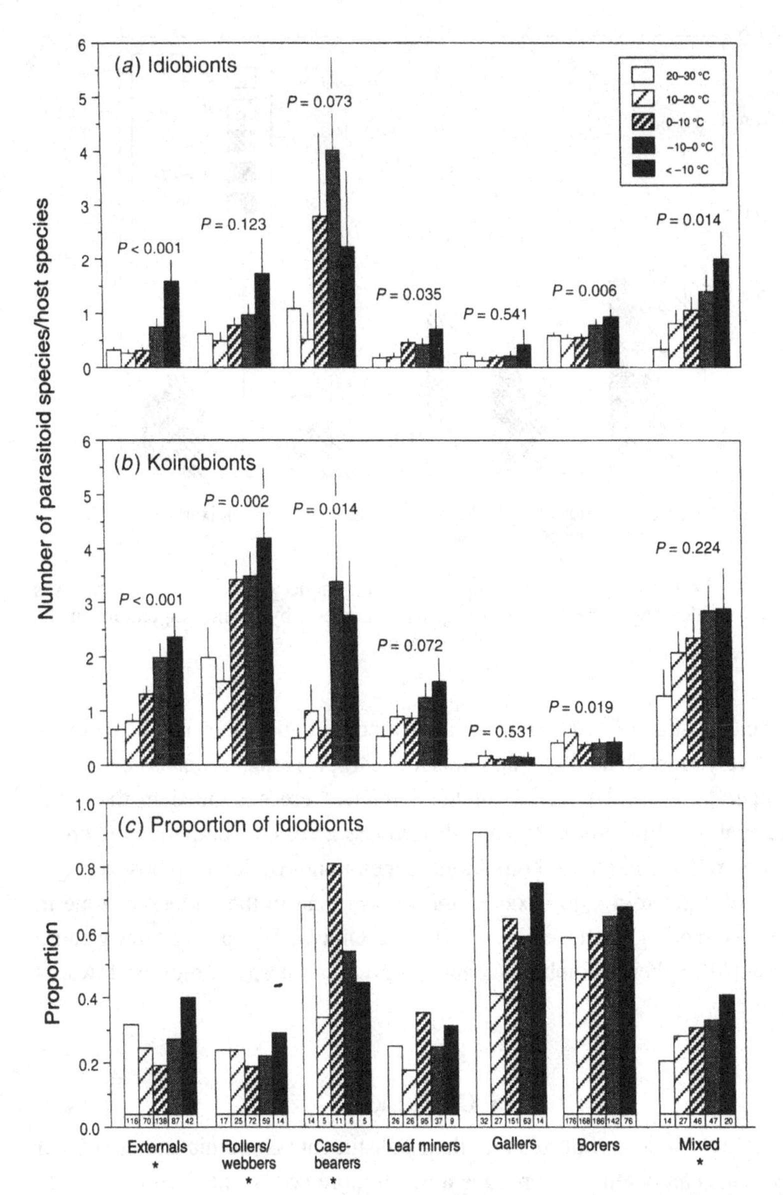

Fig. 4.9. Global, climatically-based gradients (by mean temperature in the coldest month) for Ichneumonoidea: (*a*) mean number of idiobionts, (*b*) mean number of koinobionts, and (*c*) the proportion of the ichneumonoid component comprising idiobionts. Details as in Fig. 4.6.

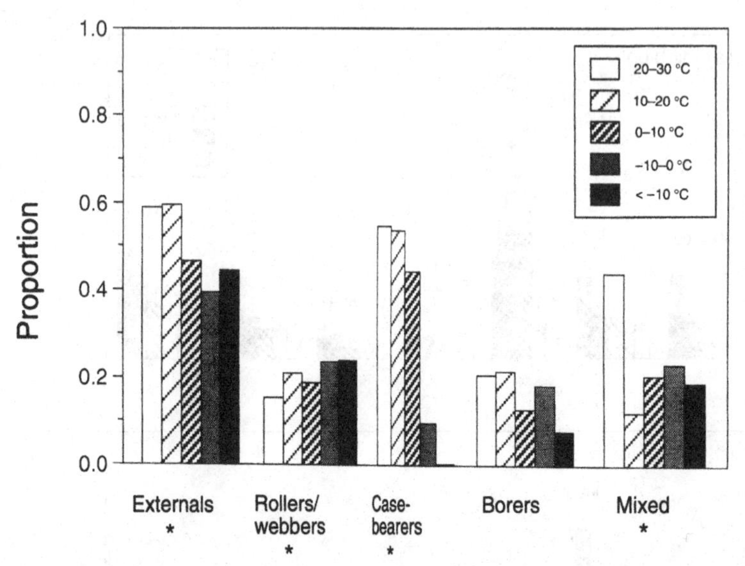

Fig. 4.10. Proportion of the koinobiont component of parasitoid complexes (Ichneumonoidea and Diptera only) comprising Diptera, by mean temperature in the coldest month.

important question of whether or not ichneumonoid idiobionts may be replaced by relatively more chalcidoid idiobionts in the tropics remains unanswered.

Temperate–tropical patterns for the Diptera were examined in the taxonomic analysis (Fig. 4.6*c*). When their richness across the climatic gradient is compared relative to that of koinobiont ichneumonoids, they compose a larger fraction of parasitoid complexes in the tropics than in the temperate zone in five of six feeding niches (Fig. 4.10). The Diptera become generally more important than their koinobiont ichneumonoid counterparts moving towards the tropics.

4.5 Conclusions

Dividing parasitoid complexes into their constituent taxonomic and biological components can potentially produce a much more complicated set of patterns than that found for total parasitoid species richness, but it should also permit a better evaluation of underlying mechanisms. What are some of the general mechanisms that influence not just the size but the composition of parasitoid complexes?

As would be expected based on the body size relationships between parasitoids and hosts, ichneumonoids dominate the parasitoid complexes associated with feeding niches that comprise a mix of large and small herbivores, whereas chalcidoids dominate the parasitoid complexes of feeding niches primarily composed of small herbivores. Dipterous parasitoids are constrained by a combination of host size and host concealment. The reasons for these associations between parasitoid and host groups appear so straightforward that there is little need to belabor this point.

Patterns of total parasitoid species richness arising from plant and habitat effects suggested a complex set of somewhat weak interactions among the characteristics of the plants themselves and the habitats in which they grow (Section 3.3). Analyses of the idiobiont and koinobiont components of these complexes appear equally complicated. But the most general pattern is that, for most feeding niches, idiobionts are most diverse on trees in both natural and cultivated habitats. Koinobionts are as rich, or richer, on hosts on herbs as on hosts on trees for endophytics, whereas on exophytics they tend to be richest on trees (in natural habitats at least). The most viable explanation for the idiobiont pattern is that of Askew (1980), who reasoned that greater numbers of polyphagous parasitoid species search trees because more host individuals and species are available in those habitats. Since these generalists will be able to use a large subset of the herbivore species constellations on trees, the number of parasitoid species per host species will be greater. This seems highly likely.

The patterns for koinobionts are more complicated. If koinobionts are generally specialized, then their failure to respond to plant type on endophytic hosts is consistent with Askew's (1980) mechanism. For parasitoids that search for one or at most a few host species, the presence of additional herbivore species on trees is irrelevant, and the number of herbivore species and koinobionts should therefore be uncorrelated. On the other hand, as I have already mentioned in other contexts, if each herbivore species is on average more abundant on trees than on herbs, we might expect to pick up more koinobionts on trees simply because their hosts are more abundant there. This could account for the increase in koinobionts on exophytic hosts found in natural habitats. Host abundance is even implicated as being important to koinobionts by the pattern found in cultivated habitats. Their failure to respond to plant type does not represent a reduction in species richness on cultivated trees (hosts on both cultivated and naturally occurring trees support about two koinobiont parasitoids per host species (see Figs. 4.4 and 4.5) but rather represents an increase in species richness on hosts associated with crop plants. Since cultivating herbs almost certainly elevates host densities relative to those found on natural herbs, we would expect koinobionts to be richer on crops. This is what is found. It is important to note that the analyses of koinobionts and idiobionts in natural versus cultivated habitats do not incorporate sample sizes (there are insufficient data), so the apparent increase in koinobionts on crops could simply reflect larger samples of crop pests than of exophytics in natural habitats. Even so, if sampling intensity is positively correlated with host density, the supposed biological basis for the pattern is probably real.

If host abundance is important to koinobionts on exophytic hosts as suggested above, why do they not also increase on endophytic hosts on trees (which cannot so easily be explained away by presumed differences in sampling intensity)? It is likely that this failure to respond is related to the quantitative and qualitative differences in the distributions of idiobionts and koinobionts in parasitoid complexes associated with exophytics versus endophytics (Section 3.2). On endophytic hosts, both koinobionts and idiobionts attack larvae. Facultatively hyperparasitic idiobionts attacking hosts within which koinobiont larvae are already developing could potentially have severe impacts on the latter, since koinobionts are almost always competitively inferior to idiobionts (see for example, Askew, 1975). This disadvantage is further aggravated when the host pupates in its feeding site, since koinobionts may have no alternative other than to do likewise (Askew, 1994); this exposes their cocoons to facultatively hyperparasitic idiobionts, which further reduces koinobiont survivorship. Thus, it is quite possible that on endophytic hosts an increase in idiobiont diversity on trees retards koinobiont accumulation. For this to be true, competition for hosts must be widespread and strong in parasitoid complexes. A number of workers have argued that competition is important among parasitoids attacking endophytic hosts (Force, 1970, 1980; Schröder, 1974; Ehler, 1979; Zwölfer, 1979; Hawkins & Goeden, 1984; Askew & Shaw, 1986; Hawkins, 1990), although it need not be universally so (Force, 1985; Hawkins & Compton, 1992). If competition is reasonably common and intense, koinobiont species richness may fail to respond to increases in host density on trees simply because idiobionts do respond.

Under this scenario, koinobionts would not be so constrained by idiobionts on exophytic hosts because the vast majority of the latter are restricted to pupae. Most exophytics move away from the larval feeding site to pupate, often in better concealed and protected locations (e.g. in the soil) where idiobionts may have more difficulty in finding them. This would be expected to improve koinobiont survivorship. Further, koinobionts that emerge before the host pupates, or which render the host pupa totally unsuitable to idiobionts, will not be subject to subsequent idiobiont attack, irrespective of idiobiont diversity. For these reasons, increases in the number of pupal idiobionts attacking exophytic hosts on trees would have minimal impact, and koinobionts would be free to take advantage of higher host densities.

Across all hosts, parasitoid complexes in cultivated habitats support relatively more koinobionts than idiobionts. But the differences are very slight (0.59 more koinobiont species per host species and 0.35 fewer idiobiont species) and are inconsistent among feeding niches (see Fig. 4.3). Most importantly, this simplistic comparison of habitats ignores the interactions between habitat type and plant type. These interactions are key to understanding the relationships between generalist and specialist parasitoids and have already been addressed in Sections 3.3 and 3.6, as well as above. Therefore, this pattern is by itself unlikely to be particularly meaningful.

The greatest challenge to understanding parasitoid species richness patterns is why some parasitoid groups are not more speciose in the tropics (Section 3.4). The patterns of idiobiont/koinobiont species richness across climatic gradients provide an opportunity to examine the mechanisms that have been proposed to explain latitudinal gradients of regional species richness. There are currently five major hypotheses, all of which address the relatively depauperate ichneumonid faunas found in the tropics. Briefly outlined in chronological order (the original papers should be consulted for detailed treatments), these are:

(1) Resource fragmentation (Janzen & Pond, 1975; Janzen, 1981). It is well documented that total, regional plant and herbivore diversity increases towards the tropics. All else being equal, the density of a temperate insect herbivore species should on average be higher on the less diverse flora than that of a tropical herbivore using flora that may be widely dispersed. Specialist parasitoids should find it much more difficult to locate sufficient host individuals to maintain viable populations in the tropics, unless their host finding abilities are substantially better than those of their temperate counterparts. If the latter is not the case, constraints on host finding should make it very difficult for parasitoids to remain specialized. However, if they are able to expand their host ranges sufficiently in the face of fragmented host resources, they should be better able to find enough host individuals in a local habitat to maintain their populations. This hypothesis proposes that parasitoid species richness is constrained by host abundance, and that parasitoid diversity is lower in the tropics because there are fewer specialists.

(2) Predation on hosts (Rathcke & Price, 1976). This hypothesis assumes that predation pressures are higher in the tropics than in temperate zones, and Rathcke and Price outline several lines of evidence that indicate that this may be the case. Further, if parasitization causes hosts to behave abnormally, this should lead to increased predation rates. Because tropical predation pressures are greater, parasitized tropical hosts should suffer relatively greater mortality

than extra-tropical hosts. This differential mortality leads to reduced parasitoid species richness in the tropics by reducing parasitoid reproductive success. Rathcke and Price also point out that this should force relatively more tropical parasitoids to exploit host stages less susceptible to predation (e.g. pupae) or hosts well hidden within, and protected by, plant parts.

(3) Predation on parasitoids (Gauld, 1987). This hypothesis shifts the importance of increased predation pressures in the tropics onto searching adult parasitoids rather than on parasitized hosts. It also assumes that resource fragmentation is operating. Gauld argues that as koinobionts are relatively slow flying and must spend additional time searching for scarce hosts in the tropics, they will be more exposed to predation. Idiobionts need not spend as much time searching, since they are niche specific and able to utilize a wider range of hosts they may encounter in the appropriate niche. Therefore, they should be relatively less susceptible to predators. Gauld provides examples of several classes of behavioral and morphological traits found among tropical parasitoids in support of this hypothesis, such as relatively more nocturnal koinobionts (which suffer less predation at night) and many idiobionts with apparent anti-predator defences (e.g. searching for hosts on foot coupled with rapid flight when disturbed, irritating bristles and spines, painful stings, and warning coloration).

(4) Interphyletic competition (Eggleton & Gaston, 1990). This hypothesis argues that insect parasitoids must compete for hosts with other parasitic organisms, such as nematodes and fungi. Because these other groups are probably more diverse in the tropics and are likely to be able to out-compete insect parasitoids when both have attacked a host, parasitoid diversity is reduced through competitive exclusion. Eggleton and Gaston provide very little empirical support for this hypothesis, but most of the logical chains in the hypothesis are at least plausible. Actually, this hypothesis is an extension of an argument made by Janzen (1975), who hypothesized that at least some Costa Rican seed bruchids are parasitoid-free because of the parasitoids' inability to compete against high levels of bacterial and fungal diseases.

(5) Nasty hosts (Gauld, Gaston & Janzen, 1992; Gauld & Gaston, 1994). This hypothesis rests upon a secondary hypothesis, which is that tropical plants more often contain toxic chemicals and/or contain a wider range of such chemicals. Tropical herbivores sequester these compounds during feeding, which consequently make them less suitable for parasitoids. Even when parasitoids are capable of evolving specializations to overcome the effects of nasty hosts, resource fragmentation exacerbates the problems against which these special-

ist parasitoids must contend, further restricting diversity. It is not critical to the hypothesis whether herbivores sequester chemicals simply as a way of dealing with them as they eat chemically defended plants, or if they actively sequester them in response to higher tropical predator pressure (see Blum (1992) for a discussion of the ways that insects use plant allelochemicals). The main predictions of this hypothesis are (a) the effects of noxious chemicals should have the greatest impact on the parasitoids associated with exophytic hosts (which will on average be eating the nastiest plant parts); (b) parasitoids attacking non-feeding host stages (eggs and pupae) should not be affected (but see Rothschild (1985) for evidence that any and all stages of aposematic Lepidoptera may be toxic); and (c) tropical parasitoids attacking nasty hosts should actually be relatively specialized. Gauld *et al.* (1992) also argue that tachinids should be less affected because their phylogenetic ancestors fed on dead and decomposing materials, and, therefore, they have had a long evolutionary history of feeding on foods containing toxins.

It should be clear from even these brief descriptions of the hypotheses that they are not mutually exclusive. For example, at least three of them incorporate resource fragmentation (resource fragmentation itself, predation on parasitoids and nasty hosts). Higher predation pressures in the tropics are also assumed in both predation hypotheses and could be indirectly involved in the nasty host hypothesis by selecting for increased chemical defenses in tropical herbivores. Therefore, we are faced with multiple potential mechanisms, several of which could be working in concert. This conceptual quagmire is not unique to hypotheses for parasitoid species richness gradients, since such gradients for other groups of organisms have led to a plethora of inter-related hypotheses (Pianka, 1978; Stevens, 1989; Rohde, 1992). The basic difference is that while those for other types of organism hope to explain why tropical species richness is so high, those for parasitoids question why their species richness is so low.

Given that many, or even all, of these mechanisms could be operating to some extent and that field experiments to distinguish them are virtually impossible using more than a handful of species at a time, is it possible at least to rank the hypotheses to estimate which factors may be most pervasive? The comparative approach is unable to test directly any of these mechanisms, but it can be used to determine if predictions arising from the hypotheses are consistent with the known patterns. Although all of the hypotheses in general predict relatively depauperate tropical parasitoid complexes, each also makes a unique combination of predictions in terms of the latitudinal patterns of species richness for the generalist and specialist components of parasitoid communities. The extent that the detailed patterns agree or disagree with these suites of predictions can provide a measure of how much each mechanism may contribute to the overall pattern. But before comparing the data with the hypotheses, it is important to establish exactly what each predicts.

One of the five hypotheses can be initially discounted as being unlikely, based on two simple predictions made by the hypothesis. The interphyletic competition hypothesis assumes that an increase in species richness of fungi and nematodes will lead to a drop in that of parasitoids through increased competition, although its proponents provide no evidence that members of species-rich systems necessarily compete more. Regarding fungi and nematode species richness patterns, Eggleton & Gaston (1990) note that 'the important role played by water in the biology of both groups suggests that they are likely to be more diverse in regions of high humidity and thus should increase in species richness in many tropical regions'. If all of their premises are true, this hypothesis predicts that parasitoid species richness will be higher in drier habitats (where fungal and nematode diversities are low) than in wet habitats (where their diversities are high). I have not analyzed rainfall patterns here, but an earlier analysis of the potential relationship between annual rainfall and parasitoid assemblage sizes (Hawkins, 1990) found that rainfall had minimal association with differences in parasitoid richness, and if any relationship was present at all, it was that parasitoid richness tended to be lowest in deserts. The relevant point is that even in habitats experiencing more than 2 meters of rainfall per year, parasitoid assemblages show no evidence of a drop in richness. Although it is true that parasitoids, fungi and nematodes must sometimes compete for individual hosts (Hochberg & Lawton, 1990) and that water availability may be needed to promote fungal/nematode diversity, the predicted association of reduced parasitoid richness and rainfall does not exist. Second, Eggleton and Gaston made no attempt to reconcile their hypothesis with another climatic/latitudinal pattern of parasitoid species richness reported in Hawkins (1990). I found that the parasitoid assemblages of root-feeding herbivores are just as rich in the tropics as in the extra-tropics. These are precisely the types of community where interphyletic competition might be expected to be most intense, since nematodes and fungi should both be most important in the soil environment. The hypothesis predicts strong losses of parasitoids towards the tropics in root-feeding communities, but this is not observed (see Figs. 3.15–3.19, pp. 46–50). Given the lack of evidence to support any of the tenets of this hypothesis and the failure of two of its predictions, interphyletic competition between parasitoids and fungi or nematodes does not appear to provide a general explanation for parasitoid latitudinal species richness patterns.

The remaining four hypotheses do have varying degrees of empirical evidence in their support. How can these hypotheses be distinguished and tested using parasitoid assemblage sizes? Table 4.2 summarizes the latitudinal trends

Table 4.2. *Latitudinal trends in species richness in parasitoid complexes attacking exophytic and endophytic hosts predicted by four hypotheses. Trends are those expected for specialist and generalist parasitoids moving from the temperate to tropical zones. Weak or strong trends are those expected for each parasitoid/host combination relative to other parasitoid/host groups*

	Exophytic hosts		Endophytic hosts	
Hypothesis	Specialists	Generalists	Specialists	Generalists
Resource fragmentation	Strong decrease	Weak decrease to increase	Strong decrease	Weak decrease to increase
Predation on hosts	Strong decrease	Strong decrease	Weak decrease to increase	Weak decrease to increase
Predation on parasitoids	Strong decrease	Weak decrease	Very strong decrease	Weak to strong decrease
Nasty hosts	Weak to strong decrease	Very strong decrease	Weak decrease to increase	Weak decrease to increase

for generalist and specialist parasitoids on both exophytic and endophytic hosts expected under each hypothesis. Because the magnitudes of the proposed forces at different latitudes are for the most part unknown, it is not possible to predict absolute changes in species richness for each group of parasitoids. But each hypothesis does lead to predictions in the richness of each parasitoid/host type combination relative to other parasitoid groups. Most of the data linking host range with parasitoid biology are for the Hymenoptera and so, at least initially, trends for specialists and generalists refer to those expected for hymenopterous koinobionts and idiobionts only. Patterns for the Diptera are important for further distinguishing hypotheses, and they can apparently be classified as generalists, but the developmental strategies they use to overcome host defenses are so different from those used by the Hymenoptera that I believe they should be examined separately.

The resource fragmentation hypothesis makes the simple prediction that specialist tropical parasitoids should be most affected, and this should be more-or-less true for specialists attacking both folivores and borers (Table 4.2). The extent that generalists are affected depends on their ability to compensate by expanding their host ranges in the tropics, but under all scenarios, they should be less affected than specialists.

Predation on parasitized hosts should be greatest on exophytic hosts, since they will be most exposed to general predation, and both specialist and generalist parasitoids would be expected to suffer under this hypothesis. It could be argued that generalists might actually suffer the most, since one possible outcome of host specialization is that parasitoids are better able to manipulate host behavior to minimize predation, which at least some hymenopterous koinobionts are capable of doing (e.g. Stamp, 1981). Even so, all types of parasitoid should be strongly affected by predation. A latitudinal gradient in predation of endophytic hosts could also be present, but it would almost certainly be weaker than for exophytics, so their parasitoids could either show a slight depression in species richness (if the gradient exists) or even an increase (if there is no such gradient).

Predation on female parasitoids while searching for hosts makes a quite distinctive set of predictions (Table 4.2). Specialists must on average spend more time searching for hosts than generalists, and so they should suffer the greatest predation. If some of this predation also occurs while parasitoids are handling hosts, parasitoids attacking endophytic hosts should also suffer high predation. Oviposition on an exposed host can be very rapid, whereas drilling a plant stem must typically take much longer. A parasitoid with its ovipositor inserted into plant tissue will be defenseless against predator attack, and it is likely that generalists will be as susceptible as specialists. But because specialists of endophytic hosts will presumably be exposed to more predation during both the searching and oviposition phases, they are likely to be the group of parasitoids most susceptible to the effects of predation.

Finally, the nasty host hypothesis predicts that herbivores that eat the most toxic plant parts should be the least susceptible to parasitoids (Table 4.2). Given that leaves are usually heavily defended chemically, particularly in the tropics (Coley & Aide, 1991), folivores might be expected to encounter noxious plant allelochemicals most often. Folivores may also use these plant toxins to defend themselves from predators, or even parasitoids for that matter, although the hypothesis as stated assumes that parasitoids are responding to chemical sequestration by herbivores rather than stimulating it. Further, overcoming these noxious chemicals will require that parasitoids become relatively more specialized. It follows that generalists attacking exophytic hosts should be the most strongly affected parasitoids, but that specialists will also suffer (partly due to the chemicals and partly due to resource fragmentation). Parasitoids of herbivores eating less toxic plant parts (such as stem borers) or that are able to manipulate plant chemistry and divert toxins from the tissues on which the herbivores actually feed into surrounding tissues (such as gallers) should be relatively less affected irrespective of whether they are specialists or generalists.

How do the real patterns match these predictions? To evaluate latitudinal gradients in idiobiont and koinobiont richness for all parasitoid taxa, only parasitoid complexes within North America and Europe provide complete data (Fig. 4.8). To extend the comparisons into the tropics, only patterns for the Ichneumonoidea (Fig. 4.9) and Diptera (Fig. 4.6*c*) are available.

To facilitate the comparison of the data with the predictions of the hypotheses, the overall patterns for idiobionts, hymenopterous koinobionts and dipterous koinobionts on exophytic versus endophytic hosts are shown in Fig. 4.11. In the northern temperate to subtropical zones, both idiobiont and koinobiont Hymenoptera richness falls towards the tropics on exophytic hosts. On endophytic hosts, significant differences are found for both parasitoid groups, but neither decreases in species richness towards the south. Either the host predation or nasty host hypothesis can account for this combination of patterns (Table 4.2).

To further distinguish these two hypotheses, it is necessary to incorporate the predicted patterns for Diptera. The expectation of Gauld *et al.* (1992) that dipterous parasitoids should be less susceptible to the effects of toxic plant chemicals and should be able to continue to utilize hosts of increasing nastiness leads to the prediction that they should either maintain their richness or perhaps fall slightly. In contrast, the host predation hypothesis predicts that they should be just as affected by predation on exophytic hosts as hymenopter-

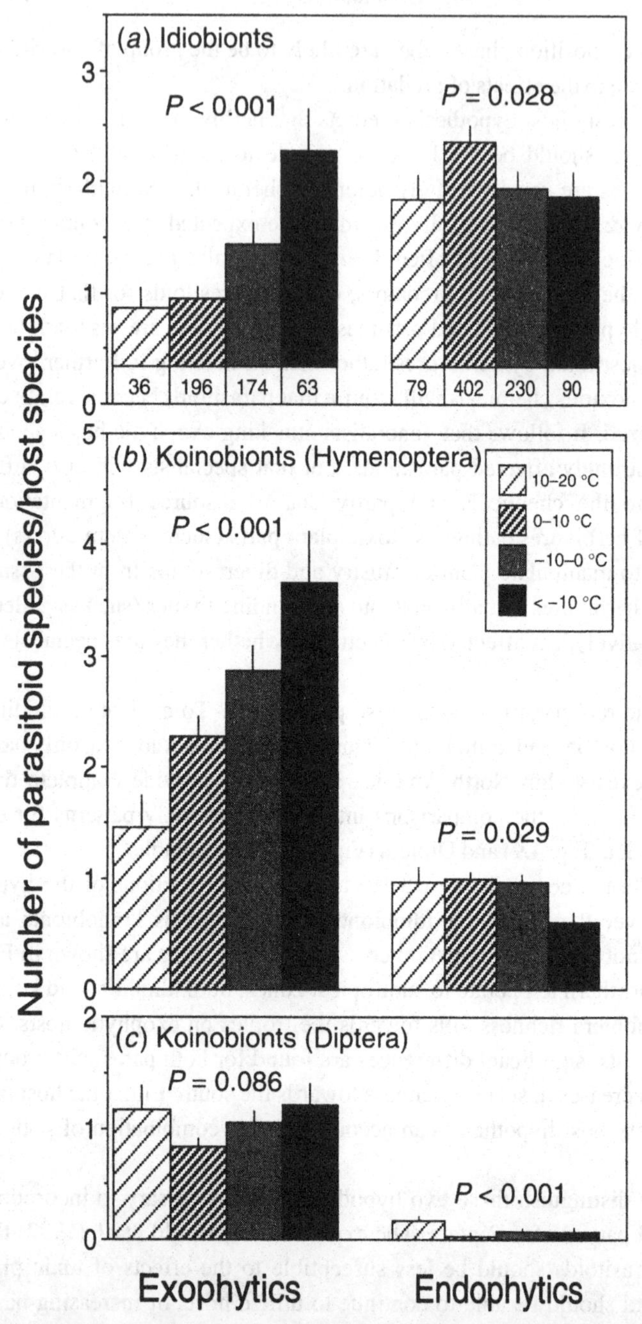

Fig. 4.11. Mean numbers of (*a*) idiobionts, (*b*) hymenopterous koinobionts, and (*c*) dipterous koinobionts composing Holarctic parasitoid complexes, by mean temperature in the coldest month. 'Exophytics' include external folivores, leaf rollers/webbers, casebearers and mixed feeders, and 'endophytics' include leaf miners, gallers, borers, and root feeders. Other details as in Fig. 4.1.

ous parasitoids, and therefore their richness should fall as severely as the Hymenoptera. The observed pattern is that they do maintain their average species richness on exophytic hosts towards the tropics (Fig. 4.11). Therefore, this group best supports the nasty host hypothesis.

Comparison of the global patterns with the predictions is less clear-cut than those for the Holarctic, at least partly because only the Ichneumonoidea and Diptera can be examined. Both idiobiont and koinobiont ichneumonoid numbers fall strongly into the tropics on externally feeding herbivores, but there are also reductions in both parasitoid groups among some of the endophytic feeding niches (Fig. 4.9). Large decreases of all ichneumonoids on exophytics coupled with smaller decreases on endophytics could be explained by either resource fragmentation, predation on hosts, or nasty hosts, so the patterns within this parasitoid superfamily are insufficient to distinguish three of the four hypotheses.

Finally, the global pattern for the Diptera adds an additional twist. Although they maintain their richness within the temperate zone, they do drop in richness into the tropics. Any of the proposed mechanisms could account for this.

In sum, comparisons of the temperate and global trends in species richness provide varying degrees of resolution for distinguishing the four potential mechanisms. But based on weight of evidence across all comparisons, the nasty host hypothesis receives the strongest support. Resource fragmentation *by itself* appears to be an insufficient explanation but is probably contributing in concert with plant chemistry. Predation on parasitized hosts cannot be completely ruled out but is unable to account for the patterns for temperate Diptera. Finally, predation of adult parasitoids receives very little support. Therefore, based on currently known latitudinal patterns in parasitoid community richness, the nasty host hypothesis offers the best explanation, although, for the logic of this hypothesis to apply, resource fragmentation must also be operating (otherwise parasitoids simply specialize and proliferate). Obviously, these comparisons of the patterns with the predictions can by no stretch of the imagination be considered critical tests of the hypotheses, so these conclusions, as weak as they are, must be considered tentative until more detailed, experimental field tests of the hypotheses can be conducted.

If we are concerned only with latitudinal gradients in species richness of parasitoid communities as a whole, one or more of the above mechanisms may be sufficient to explain the pattern. But, there remains an important aspect of the problem that is unresolved. The losses of parasitoids in tropical communities found among exophytic hosts is due to strong losses of Ichneumonoidea and weaker losses of Diptera, but chalcidoids appear at least as rich in tropical communities as in temperate ones (Fig. 4.6). If ecological processes are responsible for latitudinal trends, why are chalcidoids seemingly resistant to nasty hosts, resource fragmentation, or predation pressures?

There are at least two possibilities, one of which is not really relevant. Morrison, Auerbach & McCoy (1979) and Hespenheide (1979) argued that the absence of a latitudinal species richness gradient may be true for ichneumonoids, but may not be for chalcidoids. Fragmented or chemically defended host resources in the tropics should select for parasitoids which attack either the most abundant or least defended host stages (i.e. eggs). Chalcidoids, because they are small, will be able to attack eggs, whereas ichneumonoids will not. Morrison *et al.* and Hespenheide predicted that chalcidoid groups attacking eggs should be relatively rich in the tropics based on resource fragmentation, and Gauld *et al.* (1992) made the same prediction based on the nasty host hypothesis. That may be correct, but neither hypothesis is sufficient to explain the patterns of community richness in my data, since only parasitoids of host larvae and to a lesser extent pupae are included. Ichneumonoid communities are relatively species poor in the tropics whereas chalcidoid communities are relatively rich, and both are associated with the same host stages. The clear implication is that chalcidoids attacking all host stages are richer in the tropics. If ichneumonoid regional richness does not increase towards the tropics as rapidly as host regional richness, individual ichneumonoid communities lose species richness as the parasitoid species are diffused among more host species. Because chalcidoids do maintain their richness in the face of increasing host richness, it is likely that their richness is keeping pace with that of their hosts. If so, it is not only egg parasitoids that are richer in the tropics, and there must be some additional reason why chalcidoid latitudinal species gradients differ from those of ichneumonoids. Whatever the reason is, it seems extremely likely that workers who have claimed that the failure of parasitoids to be more diverse in the tropics is restricted to the ichneumonids and that chalcidoids probably are more diverse are right (Hespenheide, 1979; Morrison *et al.*, 1979; Noyes, 1989; Askew, 1990).

Chalcidoids and ichneumonoids differ most obviously in their relative sizes, and the answer must be at least partially due to these body size differences. Body sizes of both parasitoids and hosts have a large number of ecological ramifications relevant to all of the hypotheses for parasitoid species richness patterns. For example, in general, beyond some minimum size threshold the mean body size of a species is negatively correlated with its abundance (Gaston & Lawton, 1988; Lawton, 1990), although the relationship may be weak or absent in local communities (Blackburn *et al.*, 1993; Cotgreave, 1993). If resource fragmentation places an important constraint on parasitoids, it should be more severe on species utilizing larger, less abundant hosts. Since chalcidoids are able to utilize small hosts to a much greater extent than ichneumonoids, they should be less affected by fragmentation. Another possibility, although unlikely, is that smaller herbivores are not under the same predation pressures as large herbivores, because they are less profitable food items or perhaps because they are more difficult for predators to find. If so, small parasitized hosts may not suffer as much predation as larger ones. This reasoning could apply to adult parasitoids as well. Ichneumonoids may be attacked by a wide range of predators, whereas most larger predators (birds, lizards, etc.) may ignore chalcidoids. It is also possible that smaller herbivores do not defend themselves from predators chemically as often as larger herbivores, depending more on crypsis. Finally, chalcidoids may more commonly be gregarious than ichneumonoids since a host of any given size can potentially support many more small parasitoids than large ones. Gregarious parasitoids should be less susceptible to extinction since finding a single host individual may be sufficient for a parasitoid to avoid local extinction. Unfortunately, my data do not permit detailed tests of these speculations, but I can determine if host/parasitoid body sizes interact with species richness patterns across the climatic gradient.

Species of lepidopteran families conventionally referred to as the 'Macrolepidoptera' will be larger on average than species of 'Microlepidoptera'. The species richness of the parasitoids of these two groups can be used as an indication of the inherent suitability of large versus small host species for both ichneumonoids and chalcidoids, and then trends in latitudinal species richness for both parasitoid groups can be compared on each type of host. Because the host predation and nasty host hypotheses both presume that effects should be strongest on exophytic herbivores (and because the exophytic feeding niches contain large numbers of both large and small host species), the analysis is restricted to externals, leaf rollers/webbers and mixed exo-/endophytics. These three niches also show similar patterns of latitudinal species richness, with ichneumonoid richness falling towards the tropics and chalcidoid richness similar everywhere (Fig. 4.6), and so can be considered in concert without introducing any systematic biases. In this subset of the data, the feeding niches were pooled and range in annual temperature was used as the proxy variable for latitude.

First, Microlepidoptera support more species of Ichneumonoidea and Chalcidoidea than Macrolepidoptera (ichneumonoids: 4.44 and 2.31 species per host species, respectively; chalcidoids: 0.87 and 0.51 species per host species, respectively; Mann–Whitney *U*-tests, $P < 0.001$ for both parasitoid taxa). Second, ichneumonoids show reductions in species richness into the tropics on both types of host, but the rate of loss appears greater on large hosts

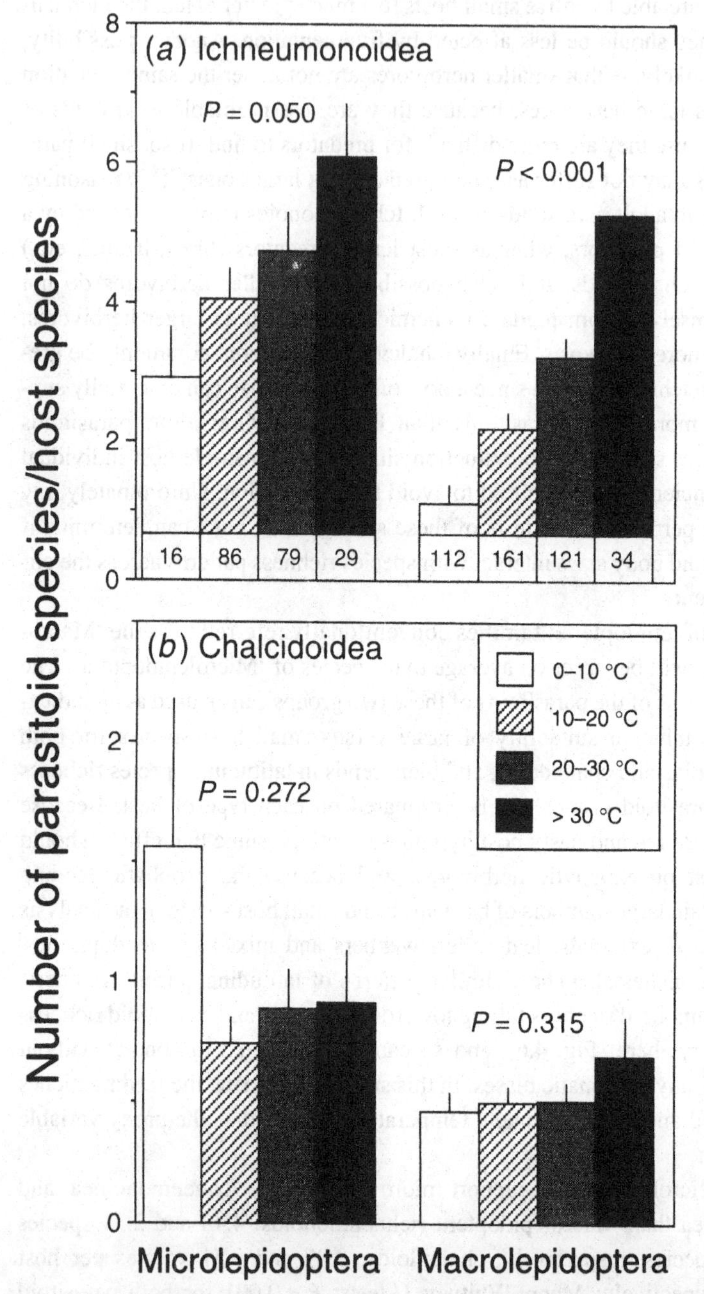

Fig. 4.12. Interactions among host body sizes (microlepidoptera vs macrolepidoptera) and the species richness of (*a*) Ichneumonoidea and (*b*) Chalcidoidea by range in annual temperature. Other details as in Fig. 4.1.

(Fig. 4.12*a*). Chalcidoids show no significant climatically based differences in either type of host, but they are richer on small hosts than on large hosts in the tropics to a much higher degree than in the extra-tropics (Fig. 4.12*b*). It appears that Macrolepidoptera are inherently poorer quality hosts everywhere, due perhaps to smaller population sizes relative to Microlepidoptera. Further, moving into the tropics they become even more unsuitable to ichneumonoids, but not to chalcidoids. As already mentioned, one possibility is that Macrolepidoptera must use chemical defenses more often to avoid general predators in all parts of the world (which also affects parasitoids), and this becomes even more important in the tropics, whereas Microlepidoptera can use crypsis everywhere (which is probably a less effective defense against parasitoids). But, if so, why are tropical chalcidoids attacking Macrolepidoptera not affected also? Alternatively, resource fragmentation may be operating for both types of host, causing the most severe losses in ichneumonoid species richness on less abundant, larger hosts. But again, if true, chalcidoids seem resistant. Are gregarious chalcidoids much more common in the tropics, such that more species can specialize on nasty hosts and/or still persist in the face of fragmented resources? Although the current body-size data are unable to resolve the possibilities, at the very least it appears that herbivore body sizes do show significant relationships with parasitoid community richness and may play a role in latitudinal species richness gradients; future analyses should take body size relationships of both hosts and parasitoids into account. Of particular interest is knowing if small hosts are relatively richer than large hosts in the tropics, since it is the former that provide the resources most commonly exploited by chalcidoids but not by ichneumonoids.

4.6 Summary

Comparing the species richness patterns of ichneumonoids and chalcidoids (the superfamilies representing the bulk of hymenopterous parasitoids) the former are richest on large-bodied hosts, whereas the latter are richest on small-bodied hosts. This straight-forward pattern reflects average body size differences between the two parasitoid groups. Further, idiobionts (i.e. parasitoids that permanently paralyze or kill their hosts during oviposition) dominate the parasitoid complexes of endophytic herbivores, whereas koinobionts (i.e. parasitoids that permit continued feeding and development by the host following parasitization) dominate the complexes of exophytic herbivores. The importance of concealment to protect otherwise defenseless hosts and idiobiont parasitoid larvae from general predators, scavengers, and weather is sufficient to account for the pattern. If parasitoid biology is associated with host range, as indicated by the bulk of the available evidence, it follows that the parasitoid complexes of endophytic/concealed hosts are dominated by generalist parasitoids, whereas those of exophytic/exposed hosts comprise more relatively specialized parasitoids.

Idiobiont species richness generally increases along the series monocots—herbs—shrubs—trees, which can be explained by the hypothesis that generalist parasitoids are attracted to larger plants supporting a more diverse and abundant herbivore fauna. Koinobionts on exophytic hosts also tend to be richest on trees in natural habitats but on endophytic hosts either do not respond to plant type or are less rich on trees. It is possible that the presence of greater numbers of competitively superior idiobionts sharing larval hosts with koinobionts, or parasitizing the koinobionts' cocoon or pupal stages, retards koinobiont accumulation on endophytic hosts on trees, but because idiobionts are largely restricted to host pupae for exophytics, larvae-attacking koinobionts are free to respond to the greater host abundances on trees. In cultivated habitats, koinobionts attacking folivores are equally rich on all types of plant, perhaps because a host abundance gradient is weak or absent.

Ichneumonoid species richness falls towards the tropics on most types of host, a result consistent with known regional patterns of diversity for this superfamily. Diptera also show a reduction in richness into the tropics, but the gradient is weaker than that for the ichneumonoids. In contrast, chalcidoid species richness per host species is as great or greater in the tropics as in the extra-tropics.

Comparisons of the latitudinal gradient of the idiobiont and koinobiont components of Holarctic parasitoid complexes against the predictions of the hypotheses that have been proposed to account for parasitoid latitudinal gradients identify the 'nasty host hypothesis' as the best available hypothesis. Resource fragmentation, the most widely cited explanation, appears unable to account for the patterns by itself, but it is likely that it is working in concert with nasty hosts. On the other hand, and importantly, a difference in the gradients for ichneumonoids and chalcidoids complicates the issue and suggests that there may be no simple explanation of latitudinal gradients in parasitoid diversity. It is possible to 'explain away' some of the discrepancies between the hypotheses that might account for a loss in tropical ichneumonid species richness and the apparent increase in tropical chalcidoid species richness, but there is insufficient information available to evaluate the possibilities. I suggest that host–parasitoid body size relationships require more attention if we are ever going to identify the mechanisms underpinning latitudinal gradients in parasitoid diversity.

5
Host mortality and parasitoid impact

5.1 Introduction

With a few exceptions involving tachinids (Edelsten, 1933; DeVries, 1984; English-Loeb, Karban & Brody, 1990), parasitoids of immature holometabolous insects invariably kill their hosts by direct parasitism or host feeding. This produces an obvious and universally recognized potential for parasitoid assemblages to cause significant levels of mortality to host populations. In a recent compilation of published life tables, H. V. Cornell & B. A. Hawkins (unpublished data) tabulated the three most important identified factors causing generational mortality for 123 species of holometabolous herbivores and found that parasitoids were listed more than any other factors, representing 89 of the pooled total of 289 (30.8%). Also, the practice of introducing parasitoids for the biological control of insect pests provides evidence that they can significantly reduce host densities, and, although not well documented in most cases, parasitoids are capable of causing sufficient host mortality to reduce host densities by one or more orders of magnitude (Embree, 1974; Beddington, Free & Lawton, 1978). On the other hand, other herbivores appear virtually immune to biological control by parasitoids (Clausen, 1978), and species are known that, at least locally, suffer minimal or no parasitoid-induced mortality (see for example, Milne, 1963; Clausen, Clancy & Chock, 1965; Wilson, 1968; Ahmad, 1974).

Despite the potential importance of parasitoids to the densities of their hosts, documenting their impact is a time-consuming task, and even when done properly, the evidence is often equivocal (Price, 1987). But despite the problems with the data and their interpretation, most workers would agree that parasitoids are sometimes important and sometimes not. The current challenge is to understand why parasitoids are important in some systems and why they are not in others (Hawkins, 1992).

Clearly, there is a great deal of variability in the amount of mortality that parasitoids inflict and its subsequent impact on host densities. Are there any identifiable patterns in host mortality and/or parasitoid impact? In this chapter I examine the factors already found to be associated with parasitoid species richness to determine if they similarly affect mortality patterns and observed reductions in host densities. Mortality patterns are analyzed using the maximum parasitism rates available for 819 herbivore species. Parasitoid impact was analyzed using the BIOCAT biological control data base, in which the proportions of introductions against different types of host that have resulted in observable reductions in host densities were used as the dependent variable.

The analytical approach is identical to that used in the previous chapters, except that following transformation the mortality data were sufficiently normally distributed and variances homoscedastistic enough to permit full factorial parametric analyses (that is, it was not necessary to examine each host feeding niche separately when analyzing multiple factors). Mortalities are presented as percentage parasitism, back-transformed following analyses of angular-transformed data. Biological control success rates were analyzed as untransformed proportions.

5.2 Host feeding niche

The feeding niche of the host has already been shown to be of primary importance in constraining the number and kind of parasitoids that attack it. Feeding niche is similarly related to host mortality, producing a dome-shaped pattern (Fig. 5.1) which corresponds closely with that for parasitoid species richness (*cf.* Fig. 3.1, p. 25). Indeed, mean maximum parasitism rate is positively correlated with mean species richness across the feeding niches (r = 0.889, F = 15.01, P = 0.018, n = 8). On the other hand, even when host feeding niche is ignored and species richness and mortality are compared directly for each host species, the association is positive and highly significant (r = 0.307, F = 85.30, P < 0.001, n = 819). Finally, when richness and mortality are compared within each feeding niche separately, correlations are significant in five cases (externals: r = 0.279, F = 15.24, P < 0.001; gallers: r = 0.254, F = 7.60, P = 0.007; borers: r = 0.181, F = 8.515, P = 0.004; root feeders: r = 0.487, F = 7.458, P = 0.012; and mixed species: r = 0.385, F = 14.23, P < 0.001), and non-significant in three, although all coefficients are positive (rollers/webbers: r = 0.154, F = 1.783, P = 0.186; leaf miners: r = 0.164, F = 1.848, P = 0.179; and casebearers: r = 0.248, F = 1.047, P = 0.321). Overall, host feeding niche is associated with both the number of parasitoids associated with a host and the total amount of parasitoid-induced mortality, and, further, hosts within each niche that support richer parasitoid complexes also tend to suffer greater mor-

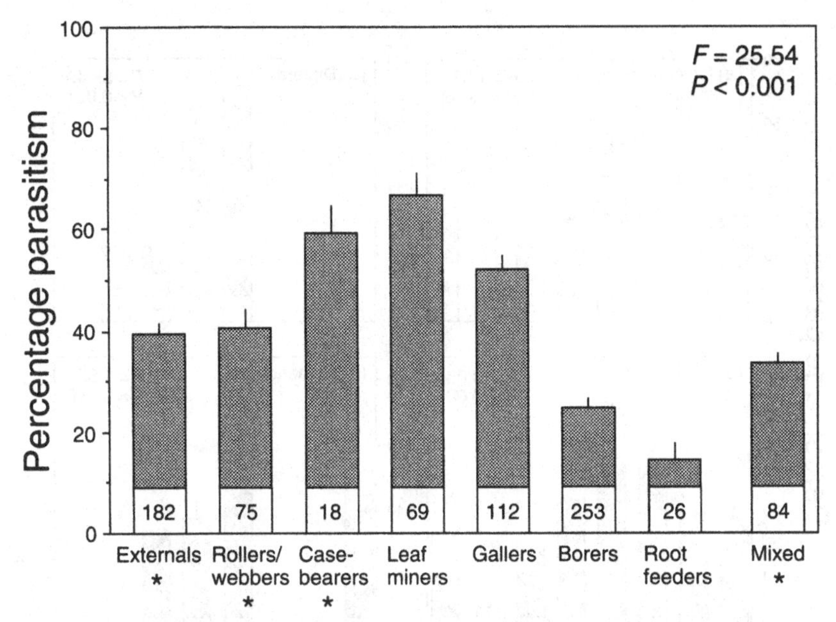

Fig. 5.1. Relationship between host feeding niche and mean maximum parasitism rate. Numbers at the base of bars are the number of host species in each niche; vertical lines are +1 S.E.M. The first seven niches are ranked in terms of decreasing motility and increasing concealment in foodplant or soil. Asterisks indicate those niches comprising completely or partially exophytically feeding herbivores.

tality. Finally, there is a relationship for individual hosts independent of feeding niche (an important result I will return to in Section 5.6).

Do all host taxonomic groups show similar patterns of mortality? Examining the relationship for each host order reveals no surprises (Fig. 5.2). Although differences in mortality among niches are not significantly different in all orders (largely due to small sample sizes), trends in all groups are consistent with the general relationship. Further, the association between species richness and mortality for individual species across all feeding niches holds within each host order (Coleoptera: $r = 0.348$, $F = 18.45$, $P < 0.001$; Diptera: $r = 0.315$, $F = 22.29$, $P < 0.001$; Lepidoptera: $r = 0.240$, $F = 25.46$, $P < 0.001$; Hymenoptera: $r = 0.403$, $F = 11.42$, $P = 0.001$). Including taxonomic groups of hosts that might be disposed toward high or low parasitoid-induced mortality does not seriously affect the nature of the general relationship. It does appear that Coleoptera often suffer lower mortality than hosts in the other orders (Fig. 5.2), but this is consistent with the general species richness–mortality relationship since beetles also support fewer parasitoids (see Fig. 3.7*a*, p. 31).

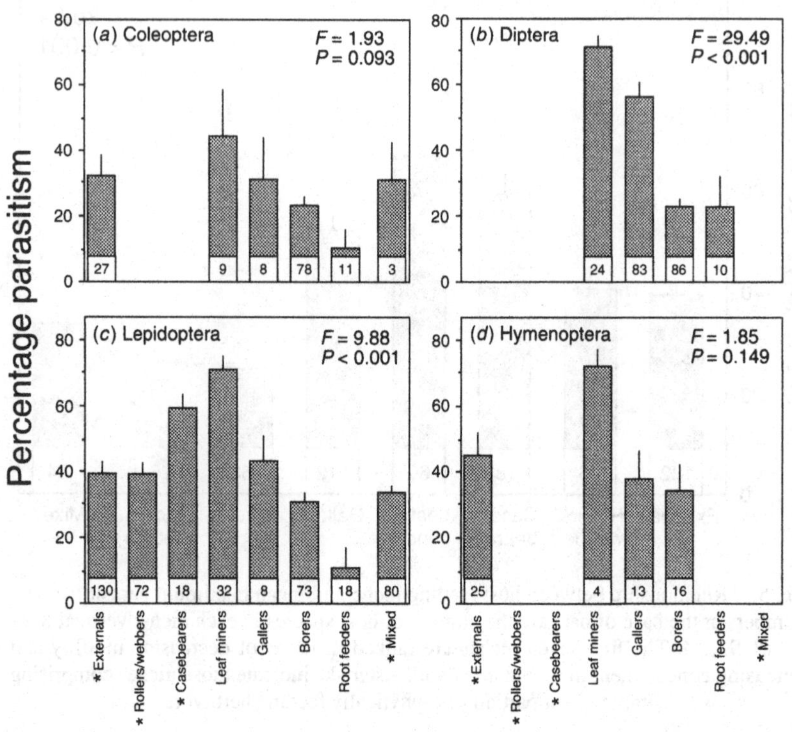

Fig. 5.2. Relationship between host feeding niche and mean maximum parasitism rate for each host order: (*a*) Coleoptera, (*b*) Diptera, (*c*) Lepidoptera, and (*d*) Hymenoptera. Details as in Fig. 5.1.

5.3 Foodplant/habitat

Herbivores on trees support more parasitoid species than those on herbs in at least some feeding niches (Section 3.3). Because parasitoid richness is positively correlated with host mortality, hosts on trees might also suffer greater parasitoid-induced mortality than hosts on herbs. Further, the effects of habitat manipulation on species richness have been shown to be generally unimportant, although complex interactions with plant type exist. Recalling that cultivation increases parasitoid species richness on herbs but decreases it on trees (see Fig. 3.14, p. 42), are patterns of mortality similarly affected?

Parasitoid-induced mortality within feeding niches does not vary significantly with regard to either plant type or habitat type (Fig. 5.3). Although hosts on trees may often support more parasitoids, they do not generally appear to suffer any greater mortality (Fig. 5.3*a*). Habitat disturbance and simplification

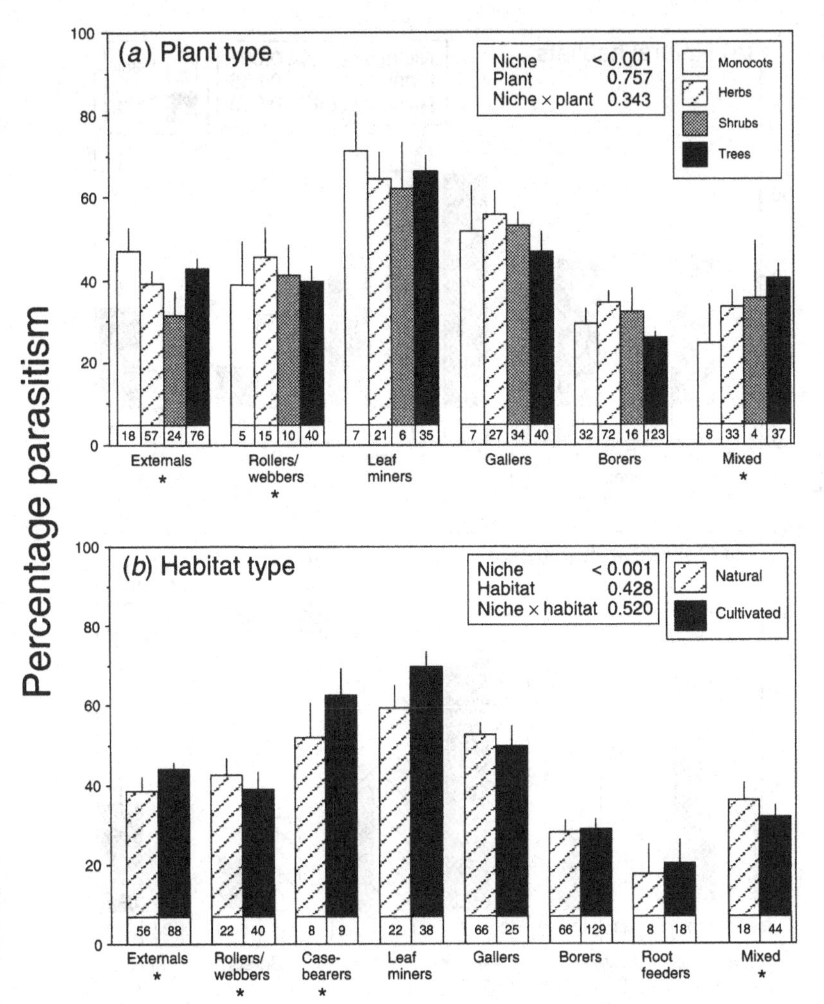

Fig. 5.3. Relationship between mean maximum parasitism rate and (*a*) host-foodplant type and (*b*) habitat type. Details as in Fig. 5.1. Numbers in the boxes represent probability values for the main effects and interactions from two-way ANOVA.

also do not reduce parasitoid-induced mortality overall (Fig. 5.3*b*). Even so, comparing mortality on herbs versus trees in natural and cultivated habitats separately does reveal an intriguing interaction. In natural habitats (Fig. 5.4*a*), plant effects are not by themselves significant, but they do interact with niche to provide a shifting pattern of mortality. Exophytic and leaf-mining hosts suffer greater mortality on trees than on herbs, mixed exo-/endophytic hosts suffer equal mortality, and gallers and borers suffer reduced mortality on trees. In

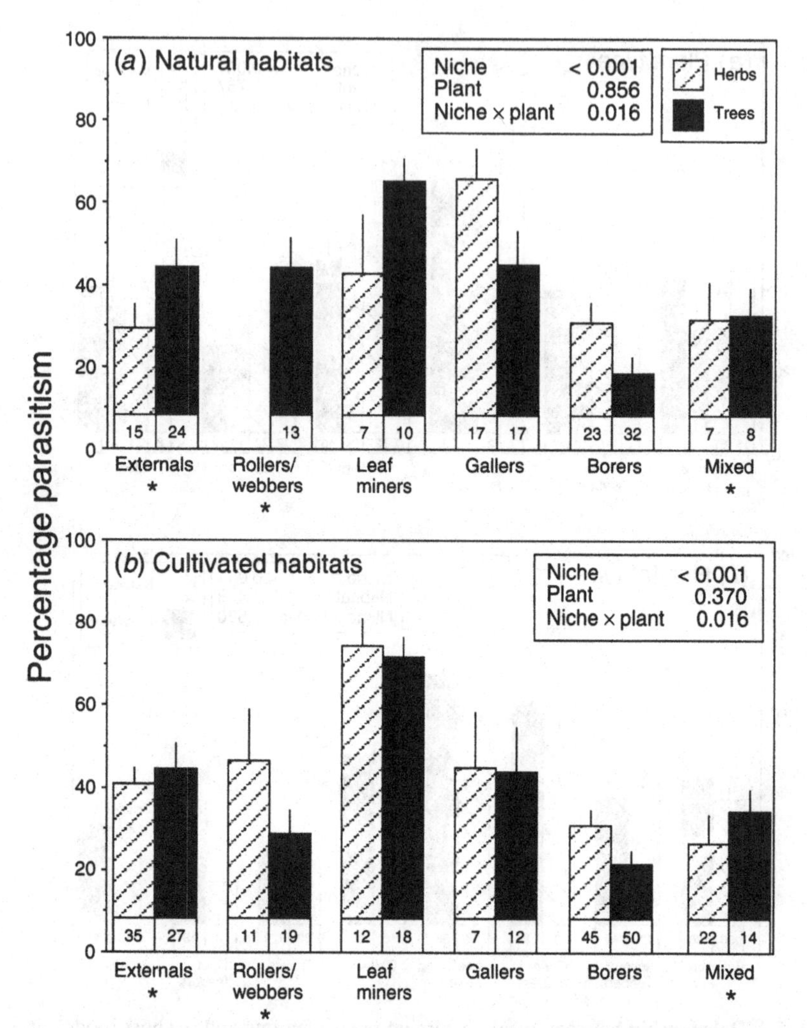

Fig. 5.4. Comparison of mean maximum parasitism rates for hosts on herbs and trees in (*a*) natural and (*b*) cultivated habitats. Details as in Fig. 5.1. Numbers in the boxes represent probability values for the main effects and interactions from two-way ANOVA.

most of the analyses up to this point, dichotomous patterns have emerged based on whether hosts are exophytic (either completely or partially) or endophytic (or underground). In this case the pattern seems to reflect more the extent that hosts are subject to parasitoid attack rather than plant position *per se*. The feeding niches can be ranked in terms of the levels of mortality (Fig. 5.1) as: leaf miners > externals > mixed > gallers > borers. The proportionate change in mortality between herbs and trees for each niche (Fig. 5.4*a*) produces an identical ranking: leaf miners (1.535) > externals (1.534) > mixed (1.021) > gallers (0.683) > borers (0.611). It appears that for hosts in the most susceptible feeding niches, the occurrence of the niche on trees exacerbates that susceptibility, whereas for hosts in less susceptible niches, the occurrence on trees reduces it.

The above applies to hosts occurring in habitats undergoing succession. In cultivated habitats, no such interactions are apparent (Fig. 5.4*b*), and hosts in each niche suffer statistically similar mortalities on herbs and trees. Therefore, if there is an interaction between feeding niche and plant type, it may apply to natural habitats only.

5.4 Latitude/climate

Are parasitoids responsible for less mortality of tropical as opposed to extra-tropical hosts as a consequence of losses in tropical species richness? Apparently not. There is no indication of a climatic gradient in mortality for any of the six feeding niches for which sufficient data are available (Fig. 5.5*a*), despite there being significant differences in the numbers of parasitoids attacking those hosts (Fig. 5.5*b*). Using range in annual temperature made no qualitative difference to the result. Therefore, the parasitoids of tropical exophytics, despite belonging to smaller assemblages, are capable of parasitizing the same proportions of host populations as their non-tropical counterparts. It also follows that tropical herbivores are just as susceptible to parasitoid attack as are non-tropical herbivores.

Interestingly, although mean species richness and mean maximum parasitism rates are clearly not associated across temperature gradients, positive correlations between the richness and parasitism of individual host species are found within all five temperature bands (20–30 °C: $r = 0.255$, $F = 9.22$, $P = 0.003$; 10–20 °C: $r = 0.278$, $F = 11.83$, $P < 0.001$; 0–10 °C: $r = 0.353$, $F = 37.67$, $P < 0.001$; –10–0 °C: $r = 0.307$, $F = 20.65$, $P < 0.001$; < –10 °C: $r = 0.323$, $F = 8.27$, $P = 0.005$). Richer parasitoid complexes tend to be associated with higher host mortality everywhere, but tropical parasitoid assemblages do relatively well irrespective of their richness.

5.5 Parasitoid impact on host densities

The above patterns indicate that richer parasitoid complexes are associated with greater mortality levels on hosts, and both dependent variables show a similar relationship with host feeding niches. We might therefore expect the

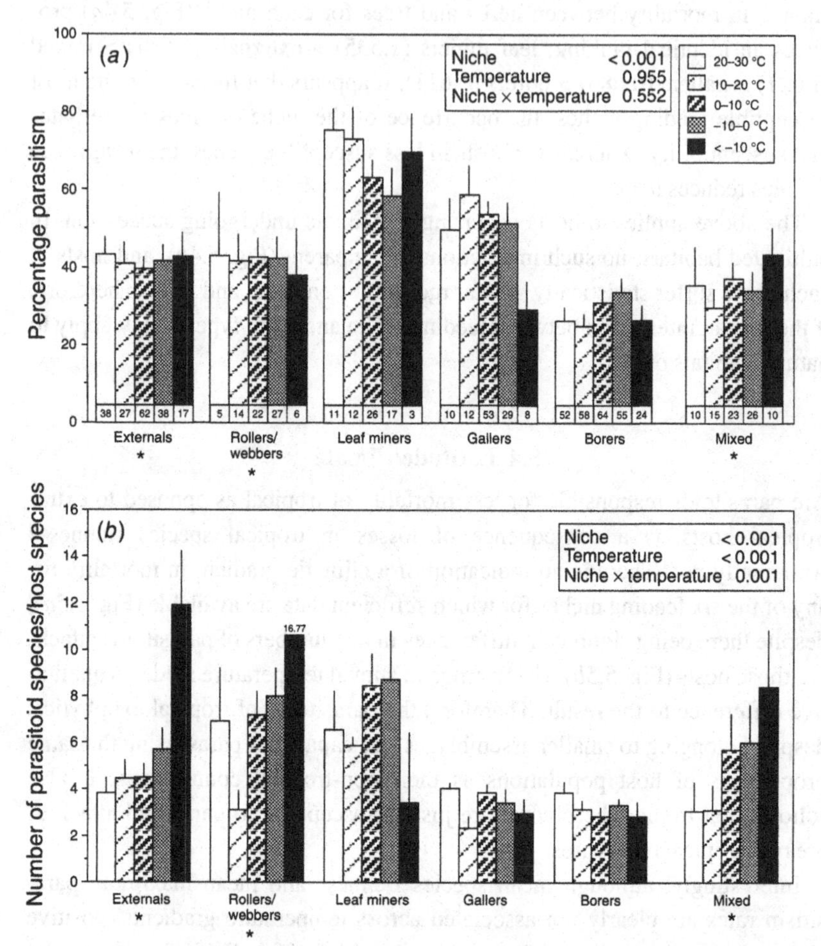

Fig. 5.5. Climatically-based gradients (using mean temperature in the coldest month) for (*a*) mean maximum parasitism rates and (*b*) mean parasitoid species richness for the hosts represented in (*a*). Other details as in Fig. 5.1. Numbers in the boxes represent probability values for the main effects and interactions from two-way ANOVA.

impact of parasitoids on host densities also to be associated with the feeding niches as a consequence of the differing susceptibilities of each to parasitoid attack, all else being equal. But all else is not equal. Herbivores die for many reasons, and parasitoids range from being the dominant cause of death to being trivial. How can the effect of parasitoids on host densities be disentangled from the effects of other mortality factors, many of which may be compensatory? Using the outcomes of biological control introductions represents one such method. Once an introduced herbivore achieves an abundance great enough to

Table 5.1. *The numbers of parasitoid introductions directed against holometabolous pests in six feeding niches with their outcomes*

Host niche	Introductions	Failures[a]	Establishments[b]	Control[c]
Externals	291	188	65	38
Leaf rollers/webbers	43	29	9	5
Leaf miners	88	51	21	16
Borers	601	429	120	52
Root feeders	146	117	19	10
Mixed exo-/endophytics	197	137	37	23
Total	1366	951	271	144

[a] Parasitoid not established.
[b] Parasitoid established but no control reported.
[c] Parasitoid established and some degree of control reported.

be considered a pest, tracking host densities after the introduction of a parasitoid imported from the host's native range should provide at least a crude measure of the impact of the parasitoid on host populations that occurs in addition to any pre-existing mortality factors (Beddington *et al.*, 1978; Hassell & Waage, 1984). Although pest reductions following parasitoid introductions have historically been poorly documented quantitatively and control populations are rarely monitored, the simple observation by biocontrol specialists that the parasitoid successfully controlled the host to some extent after its introduction probably indicates that the parasitoid had some impact. The biological control record undoubtedly contains errors and omissions, but it does represent the best available large data set for comparisons with parasitoid species richness and host mortality data. More importantly, the errors inherent in the biological control data should not necessarily be correlated with the independently generated data sets with which they can be compared.

Because of a relative paucity of data, the biological control data were treated slightly differently from the other data sets. First, gallers were excluded. They were represented by only seven introductions against two host species, a sample size judged to be insufficient to provide a reasonably reliable estimate of success rates. Second, the only records for casebearers were two *Coleophora* species (Lepidoptera: Coleophoridae), both of which are casebearing leaf miners. This was also insufficient to justify categorizing success rates for this niche, and because they are essentially leaf miners, these hosts were classed as such. Under these criteria, the final data set contains the outcomes of 1366 introductions against hosts in six feeding niches (Table 5.1). Finally, the data

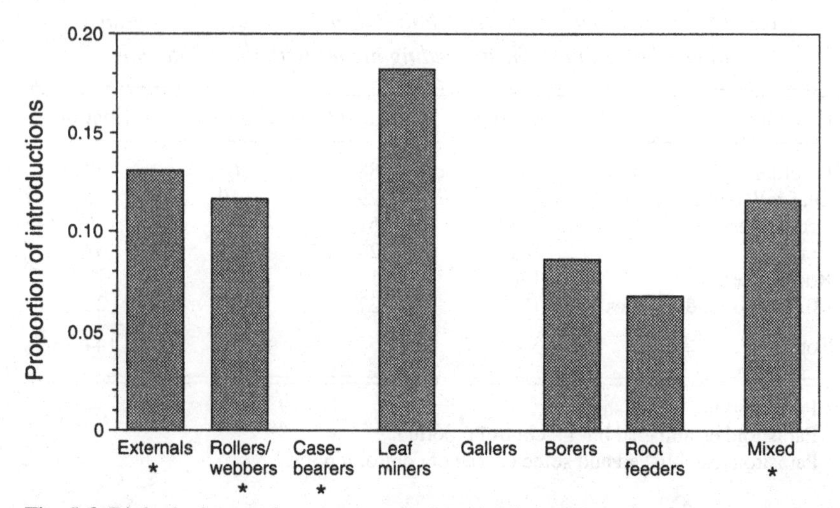

Fig. 5.6. Biological control success rate by host feeding niche measured as the pooled proportion of a parasitoid introductions against all pest species in each niche resulting in some control of the host.

were not distributed broadly enough to permit them to be subdivided into small groups for analyses of multiple factors, such as plant/habitat type or climate, and so analyses were restricted to comparisons among host feeding niches and host taxonomic groups (orders).

The pattern of variation in success rate takes a form similar to that previously found for parasitoid species richness and host mortality (Fig. 5.6). The leaf rollers/webbers are slightly out of sequence, but this proportion is based on the smallest number of parasitoid introductions and therefore is the most sensitive to small differences in the number of recorded successes. Nevertheless, positive relationships among the variables are supported statistically. Biological control success rates across the host feeding niches are correlated with both mean parasitoid species richness ($r = 0.851$, $F = 10.51$, $P = 0.032$) and mean maximum parasitism rate ($r = 0.983$, $F = 114.25$, $P < 0.001$). These results are fully consistent with the idea that characteristics of herbivores' feeding biologies broadly measure their susceptibility to parasitoid attack and link parasitoid species, parasitism rates, and the impact of parasitoids on host densities.

The above relationships are based on a wide range of host taxa. But parasitoid species richness varies among herbivore orders (see Fig. 3.7, p. 31), and both parasitoid establishment rates and biological control success rates are also known to differ among orders (Hall *et al.*, 1980; Stiling, 1990; Gross, 1991).

The biological control data are taxonomically independent of the other data sets, since the former represent successes against introduced pests, whereas the latter are based on native insects, only a few of which are also represented in the biological control data. Despite taxonomic independence among the data sets at the species level, it would be useful to be able to judge to what extent the patterns may be influenced by the inclusion of four orders of hosts that inherently support differing numbers of parasitoid species.

To examine the patterns in more detail, I calculated values for parasitoid species richness, percentage parasitism, and biological control success rates for each of the four host orders in the three data sets (see Hawkins (1993a) for the data). Subdividing the data permits more detailed comparisons, but it also generates two analytical problems. First, it creates empty cells because herbivores with certain feeding biologies are either absent in the different host orders or have not been studied. Second, it reduces sample sizes within each cell, decreasing confidence in the reliability of estimates of the means. Both problems are most severe in the biological control data set.

We have already seen that the relationship between parasitoid species richness and host mortality holds within each host order (Section 5.2). For the relationships between these variables and biological control success rates, both mean richness and mean parasitism are correlated with success in the Lepidoptera (richness: $r = 0.935$, $F = 20.86$, $n = 6$, $P = 0.020$; mortality: $r = 0.977$, $F = 63.08$, $P = 0.004$). For the Coleoptera, mean species richness is not correlated significantly with biological control success ($r = 0.664$, $F = 2.36$, $n = 5$, $P = 0.222$), but the relationship between mean parasitism and success is significant ($r = 0.891$, $F = 11.57$, $P = 0.042$). Neither correlation is significant for the Hymenoptera (richness: $r = 0.361$, $F = 0.149$, $n = 3$, $P = 0.765$; mortality: $r = 0.953$, $F = 9.862$, $P = 0.196$) or Diptera (richness: $r = 0.410$, $F = 0.40$, $n = 4$, $P = 0.590$; mortality: $r = 0.547$, $F = 0.85$, $P = 0.453$). It is noteworthy that mean mortality shows a stronger relationship with success than does mean species richness in all cases whether statistically significant or not.

It appears that, overall, similar patterns occur in all host orders. Not all correlations are statistically significant, but this may reflect an inability to detect relationships when sample sizes are small rather than an absence of relationships. When the number of feeding niches found within a host order is relatively large (i.e. Lepidoptera), the statistics are stronger. As the number of feeding niches becomes more restricted (from Coleoptera to Diptera to Hymenoptera), the statistics become weaker. At least all correlation coefficients are positive, and it appears that all host orders contribute to the general pattern.

5.6 Conclusions

The analysis of maximum percentage parasitism reveals a number of relationships that suggest that host mortality and parasitoid species richness follow many of the same rules. As with species richness, mortality is more strongly associated with feeding niche than with any of the other variables examined, and the relationships between feeding niche and both dependent variables appear strikingly similar. On the other hand, species richness is itself associated with mortality even when the effects of feeding niche are ignored by directly comparing individual host species. This raises the question as to the cause and effect relationships among feeding niche, parasitoid richness, and mortality.

The association between species richness and mortality in the feeding niches could arise for two reasons. First, it is tautological that each parasitoid species must contribute to host mortality for it to be present. It is possible that the correspondence between the variables arises simply because more parasitoid species must result in more mortality (e.g. all parasitoid species subject their hosts to some average mortality that is similar for all feeding niches, and twice as many species in a feeding niche consequently produces twice as much mortality in that niche). A second possibility is that feeding niche influences the susceptibility of hosts to parasitoid attack, and this drives both species richness and mortality, but indirectly. For example, more susceptible hosts may support many parasitoid species, but most of the mortality actually comes from relatively few parasitoids that are well adapted to exploit the hosts' susceptibility, while many of the additional parasitoids are simply opportunists that attack susceptible hosts at low rates because they are easily discovered by a few searching females. These alternative relationships between parasitoid species richness and host mortality have been termed 'causative' and 'correlative' by Towner (1992).

One way of distinguishing these alternative possibilities is to examine mortality patterns among the feeding niches after accounting for the number of parasitoid species attacking each host species. If the relationship between species richness and mortality is causative, niche effects on mortality will be less pronounced (or absent) once parasitoid richness is factored out, whereas if it is correlative, differences among the niches should persist.

The effect on percentage parasitism by adjusting for parasitoid species richness under the two potential mechanisms can be illustrated using a simple, hypothetical data set (Table 5.2) that demonstrates how ANCOVA works. The data represent four host species in each of four 'niches' ranked by average parasitoid species richness. Within the niches, host species show variation in

Table 5.2. *Hypothetical data illustrating two alternative mechanisms (causative and correlative) which can produce a positive association between parasitoid species richness and host mortality (see text for explanation)*

Host niche	Parasitoid species richness	Percentage parasitism: Causative	Percentage parasitism: Correlative
1	1	3	10
	2	7	11
	3	8	10
	4	12	11
2	5	16	21
	6	18	23
	7	22	22
	8	24	22
3	9	26	31
	10	30	32
	11	32	32
	12	37	31
4	13	39	41
	14	41	42
	15	46	43
	16	48	42

the number of parasitoid species each supports. The associated mortality data represent the alternative possibilities. In the causative case, each and every parasitoid species contributes 3% parasitism (with a small amount of 'measurement error'). In the correlative case, hosts in the four niches differ in their susceptibility to attack by a single, dominant parasitoid species by 10% such that these parasitoids are responsible for 10% parasitism in niche 1 to 40% in niche 4. All remaining parasitoids add only an additional 1–2% mortality (again with a small amount of error). Therefore, in the causative case species richness directly drives the pattern of total mortality, whereas in the correlative case species richness and mortality are associated only because both are related to niche, with differences in host susceptibility to a single, dominant parasitoid species underlying the relationship, and the remaining parasitoids contributing only a small amount of 'noise'.

In both cases, raw host mortality across the niches is strongly associated with parasitoid species richness (Fig. 5.7*a*, *b*). After adjusting for the number of parasitoid species with ANCOVA (Fig. 5.7*c*), mortality across niches becomes flat in the causative case, whereas in the correlative case the relationship with parasitoid species richness is maintained. Adjusting host mortality by

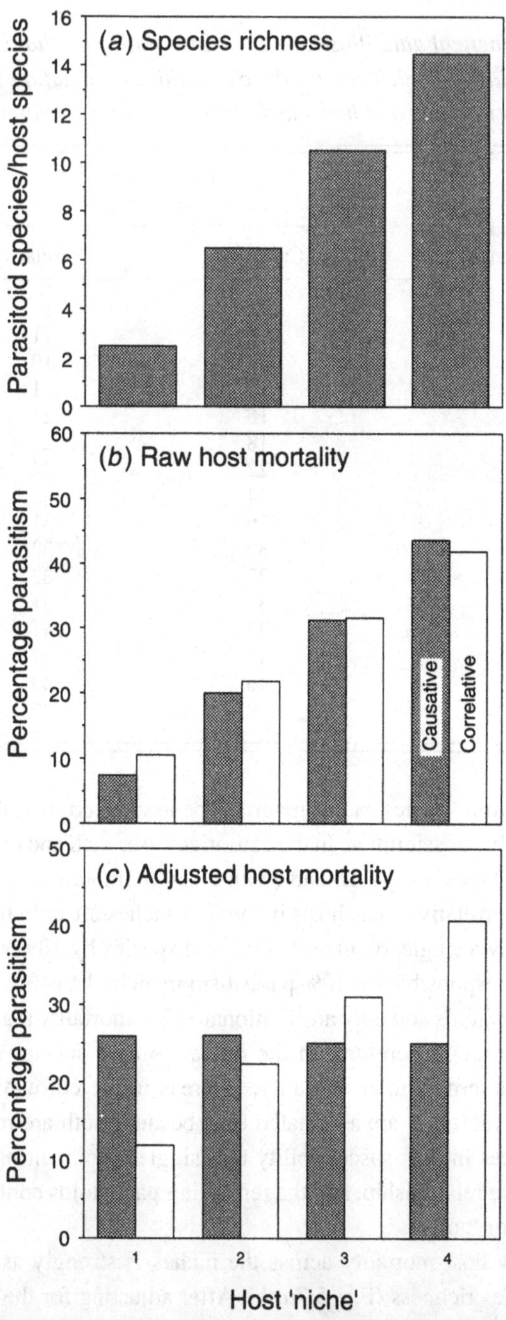

Fig. 5.7. Tests of the 'causative' and 'correlative' mechanisms underlying parasitism rates (data given in Table 5.2). (*a*) Mean species richness in the four host 'niches'. (*b*) Unadjusted percentage parasitism. (*c*) Percentage parasitism after adjusting for the number of parasitoid species. See text for further explanation.

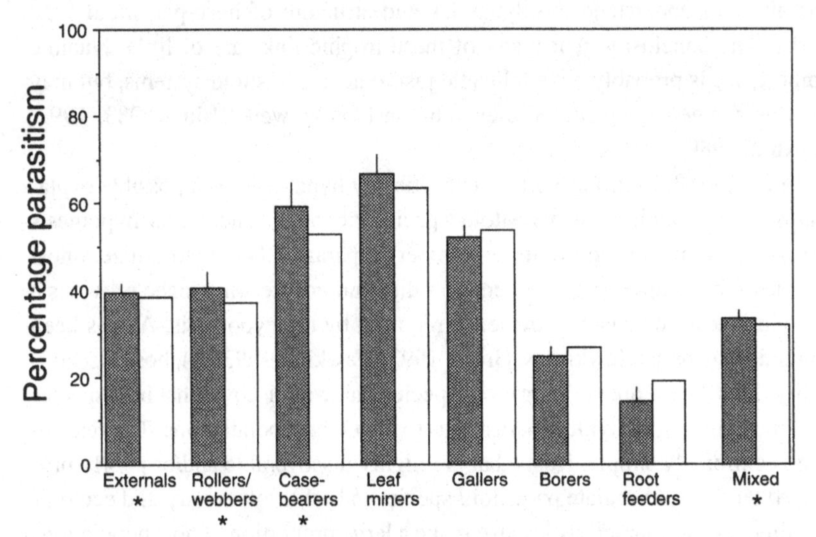

Fig. 5.8. Comparison of unadjusted parasitism rates (from Fig. 5.1) (▩) and parasitism rates after adjusting for parasitoid species richness (□).

the number of parasitoids responsible for it produces two quite distinct patterns, depending on which mechanism drives mortality.

So how does adjusting real mortality, which undoubtedly lies somewhere between the hypothetical extremes, affect its relationship with host feeding niche? The adjusted means from an ANCOVA of parasitism against feeding niche with species richness as the covariate look very similar to the non-adjusted means (Fig. 5.8) (the preliminary test for homogeneity of slopes was non-significant, $P = 0.370$). It appears that the data conform more to the correlative mechanism than to the causative mechanism, and differences among the niches do not reflect simply that more parasitoid species attacking some types of host produce more total mortality, with each contributing similar levels. Instead, hosts in the more susceptible feeding niches are subject to higher overall attack rates independently of parasitoid richness due to differing susceptibilities to one, or perhaps a few, dominant parasitoid species. It also follows that many of the additional parasitoid species associated with susceptible hosts are relatively rare, contributing very little to overall mortality. Therefore, most of the extra links in the species-rich parasitoid–host systems appear weak, with little or no relevance to host-population dynamics. A logical consequence of this result is that understanding host–parasitoid interactions may be much simpler than would be suggested by the species richness of the parasitoid communities; it may be sufficient to document the interactions of only a few key members to understand the dynamics and structure of host–parasitoid food webs. The conclusion that many of these trophic links are of little dynamic importance is probably not applicable just to host–parasitoid systems, but may be true for general predator–prey relationships as well (Paine, 1983, 1992; Lawton, 1989).

In Chapter 3, I introduced the susceptibility hypothesis as a possible explanation for variability in parasitoid species richness. Under this hypothesis refuges from parasitism limit the number of parasitoids able to utilize a host species. The relationship between mortality and host feeding niche provides a second link in the chain of evidence required by the hypothesis. As has been argued elsewhere (Hawkins & Gross, 1992; Hawkins, 1993*a,b*), host susceptibility links the number of parasitoid species that will discover that host species and the ability of individual parasitoids to attack host populations. The reasoning is extremely simple. Hosts that are highly discoverable and/or poorly protected tend to accumulate parasitoid species in both evolutionary and ecological time. These characteristics also make a large proportion of host populations vulnerable to attack by parasitoids, resulting in high parasitism rates. Although it does not necessarily follow that hosts that support rich parasitoid complexes *always* suffer high parasitoid-induced mortality, this should occur frequently enough for species richness and mortality to be correlated.

The reduction in host mortality along the series leaf miners—gallers—borers—root feeders may be readily explained by increased protection from parasitoids as hosts enclosed by plant tissue or the soil become more difficult to locate and attack. But most exophytic hosts clearly have evolved defenses against natural enemies that affect parasitoids, and, on average, smaller proportions of external folivores are killed by parasitoids than in the cases of casebearers, leaf miners, or gallers (Fig. 5.1). My data do not permit an evaluation of the effectiveness of the various defenses that exophytics may employ (see Gross (1993) for a review of these defenses), but it should be obvious that susceptibility to parasitoid attack may involve a wide range of factors, some arising from the host's foodplant and others arising from the herbivores themselves. Yet another complication with exophytics is that they are probably more exposed to general predation and weather, so larger proportions of their populations will be unavailable to parasitoids. The situation with endophytic hosts appears relatively straight-forward, and they are able to take advantage of the defenses provided by the plant parts they infest to protect themselves from an important source of mortality. Exophytic hosts, in contrast, must utilize intrinsic defenses against parasitoids (e.g. immune responses, plant allelochemicals, evasive behaviors, etc.), and the apparent effects of parasitoids on host populations will be complicated by stronger interactions among parasitoids, predators, and climate. The most that can be concluded here is that the sum of intrinsic defenses employed by exophytic folivores makes them appear less susceptible to parasitoids than are leaf miners and gallers, but more susceptible than borers or root feeders.

The third pattern that implicates host susceptibility as the most important mechanism defining host–parasitoid interactions is that host feeding niche links biocontrol success rates and host mortality. When hosts are highly susceptible to attack, at least some parasitoid species more often manage to kill large numbers of hosts, maximizing the impact of parasitoids on host densities. This is reflected in the biological control record as a greater proportion of parasitoid introductions that have led to successful control. Therefore, in addition to predicting relationships between parasitoid community species richness and parasitoid-induced host mortality, the susceptibility hypothesis may provide a tool for predicting when the introduction of parasitoids for the biological control of insect pests can at least potentially succeed. It can be argued that, in practice, parasitoid introductions may fail to successfully control pests for many reasons, such as misidentification of hosts or parasitoids, the introduction of unsuitable biotypes, poor synchronization with the host, climatic mismatch, insecticide interference, adverse crop management practices, the release of too few individual parasitoids, etc. This is all true. On the other hand, if these factors were of overriding importance, they should swamp out the feeding niche relationship in the biological control record. That the pattern exists in spite of a large number of potential confounding factors suggests that inherent host susceptibility, estimated by feeding niche, represents a potent force constraining the ability of parasitoids to have an impact on host densities.

Caution should be exercised when interpreting the feeding niche–biological control relationship. I have argued that feeding niche represents a measure of host susceptibility, but it must be borne in mind that it is an *indirect* measure. It does not follow that all leaf miners are equally susceptible to parasitoid attack whereas all root feeders are resistant. There are examples of successful biological control for all types of herbivore. This reflects the fact that parasitoids are capable of evolving responses to host refuges to maximize their own reproductive success. Two alternative, but related, parasitoid responses are possible. First, parasitoids can be 'refuge breakers', by developing behavioral and morphological characteristics that allow them to utilize hosts that otherwise would be immune to attack. For example, some tachinids, which lack plant-piercing ovipositors, attack stem-boring caterpillars by placing active larvae near host entry holes; the larvae can then crawl through the tunnel until the host is encountered (Clausen, 1940). Second, parasitoids may be 'refuge avoiders', attacking a host at a point in its life cycle when it is out of its refuge.

Platygasterids, despite their small size and short ovipositors, can achieve high parasitism rates of gall midges simply by ovipositing in host eggs before the gall has developed. The existence of refuge-breakers and refuge-avoiders makes blanket statements about the susceptibility of individual host species in the feeding niches highly risky. Although feeding niche appears to provide an average estimate of a host's refuge from parasitism, there are undoubtedly many 'exceptions'. This is because feeding niche is a characteristic of the host, whereas a refuge actually arises from the host–parasitoid interaction. A much more direct measure of the susceptibility of a particular herbivore species to parasitoid attack and the ability of parasitoids to depress host densities will be the maximum level of parasitism that the parasitoids can achieve, since this incorporates both host defenses and the parasitoids' evolutionary responses to them. If the susceptibility hypothesis is correct, maximum parasitism rate should provide a much more accurate predictor of the impact of parasitoids on host population dynamics and biological control success rates than would the host's feeding niche. I will return to this issue and prediction in Chapter 7.

Another aspect of the mortality patterns that is relevant to biological control is that parasitism is positively associated with parasitoid species richness. This could be interpreted to indicate that rich parasitoid complexes are more likely to have an impact on host populations than would depauperate ones, as I have suggested elsewhere (Hawkins, 1993b). This would support the biological control philosophy of multiple introductions (see Ehler, 1990) and would predict that biological control success rates should be positively associated with the number of parasitoid species introduced and established. This prediction receives very little empirical support in the biological control record, since it is a hallmark of most successful projects that control is achieved by one or perhaps two natural enemy species (Doutt & DeBach, 1964; Huffaker, Messenger & DeBach, 1974). However, the finding that the relationship between percentage parasitism and feeding niche is independent of species richness and appears to reflect correlation rather than causation, as discussed above, indicates that the species richness–biological control success rate relationship is a red-herring. High parasitoid species richness is not required for great impact on host densities. Host depression arises as a consequence of the host being highly susceptible to attack. It is probably no accident that percentage parasitism shows stronger statistical relationships with biological control success than parasitoid species richness; under the susceptibility hypothesis the latter relationship reflects cause and effect, whereas the former is simply correlative. Many parasitoids are not necessary for successful control; one is sufficient if it can take advantage of a host's vulnerability. This is consistent with documented biological control successes (Myers, Higgens & Kovacs, 1989).

Therefore, the patterns are not inconsistent with the principle of single introductions. Given a choice of several potential biological control agents, the species that is, by itself, capable of causing the highest parasitism rates is the preferred species to introduce. Although this conclusion is simple in theory, it remains to be seen if it is that simple in practice.

Beyond the relationship between mortality and host feeding niche, other factors appear weakly associated with host mortality. Neither plant type nor habitat type has any general significant influence. Host susceptibility does appear to be influenced by plant type in natural habitats (Fig. 5.4*a*), but not in cultivated habitats (Fig. 5.4*b*). The difference in the patterns between the two habitat types is difficult to explain fully. A possible explanation for the shifting pattern in natural habitats is that for hosts that are generally susceptible to parasitoids (leaf miners and exophytic folivores), processes generated during ecological succession increase that susceptibility as envisaged by Price (1991, 1994). For less susceptible hosts, on the other hand, parasitoids become even less effective during succession due to increasing plant structural protection. Endophytic hosts feeding deep within plant tissues receive at least some protection from parasitoid attack by virtue of the physical characteristics of the plant parts they inhabit. Parasitoids must first be able to recognize the presence of hosts boring through plant stems, for example, and once having done so must drill through the plant tissue to oviposit. It seems likely that, on average, it will be easier for a parasitoid to locate a borer and penetrate enveloping plant tissues when the host is in a small, non-woody plant than when it is in a tree trunk. Gallers could benefit from increased protection on trees as well if tree galls are on average woodier and more difficult for parasitoids to penetrate; gall toughness is known to reduce parasitism rates (e.g. Craig, 1994).

The lack of any pattern in cultivated habitats (Fig. 5.4*b*) lends further, indirect support for a succession hypothesis, since it implies that decoupling plant type from succession accounts for the absence of effects on susceptible hosts. But if plant structural protection is also operating, as suggested by the analysis of natural habitats, gallers and/or borers should remain less susceptible on trees than on herbs in cultivated habitats as well. The trend is in the right direction for borers (Fig. 5.4*b*), but galler mortality is virtually identical on both types of plant. Therefore, the analysis of natural habitats suggests that a simple succession-based mechanism is insufficient, whereas the analysis of cultivated habitats suggests that plant structural protection is also insufficient. But it should be borne in mind that all of these relationships, whether statistically significant or not, are based on a relatively limited number of cases. Data for additional host–parasitoid systems are probably necessary to judge accurately the importance of plant/habitat factors for host susceptibility to parasitoid attack. Given the limited scope of the data, I do not think that it is worth speculating too much about the plant type–habitat interactions or attempting to explain every quirk in the patterns, which very quickly moves into the realm of 'ad hochery'. Patterns predicted by theories of host–parasitoid interactions that use ecological succession as a template (Southwood, 1976; Price, 1991) are difficult to evaluate in any general sense, at least partially because I do not have enough cases to permit reasonably powerful tests.

The complete absence of a latitudinal gradient in parasitism rates is interesting for several reasons. First, it is consistent with the conclusion that the relationship between parasitoid species richness and host mortality is correlative rather than causative. A probable explanation is that although parasitoid complexes of exophytic herbivores are richer in the temperate zone than in the tropics, most host mortality is caused by a few, key parasitoids, and the greater parasitoid richness in the temperate zone has little impact on total host mortality. Second, I discussed in Chapter 4 the major mechanisms that have been proposed to account for the relatively depauperate species richness of parasitoids in the tropics. The nasty host hypothesis, underpinned by resource fragmentation, appears to offer the best general explanation for patterns of latitudinal parasitoid species richness, but it seems to make little difference to the importance of parasitoids as mortality agents in different parts of the world. Plant allelochemicals may decrease the suitability of exophytic hosts to more generalized parasitoid species (Barbosa, 1988), but they may account for relatively little host mortality anyway. If so, specialist parasitoids that are adapted to nasty hosts are capable of finding similar numbers of host individuals in both tropical and extra-tropical habitats, despite any increased difficulties that they may encounter in searching highly fragmented, tropical habitats. Finally, it appears that parasitoids play very similar roles in the dynamics of host populations in all parts of the world, in the sense that similar proportions of hosts are killed. The biological control record further suggests that this is indeed the case; there is little evidence that biocontrol is more effective in either the tropics or extra-tropics (Greathead, 1986; Hokkanen, 1986).

In sum, feeble or non-existent gradients in mortality found in the plant, habitat, and climatic analyzes suggest that beyond the inherent susceptibility of herbivores to parasitoid attack related to their feeding behavior, other larger scale factors show at most very limited relationships with parasitoid-induced host mortality.

5.7 Summary

Host feeding niche emerges as the strongest correlate of the maximum levels of mortality that parasitoids inflict on host populations. Because parasitoid species richness is similarly associated with feeding niche, host mortality could be either directly driven by parasitoid species richness, or alternatively, their relationship could be spurious. An Analysis of Covariance indicates that this association is 'correlative' rather than 'causative'; that is, niche differences in mortality levels arise from differences in host susceptibility to attack by one or a few parasitoid species rather than from a direct relationship between parasitoid species richness and total parasitism rate. If so, most of the extra parasitoid species found in species-rich systems contribute little to host mortality or population dynamics.

Host mortality shows weak relationships with plant and habitat type, although there is evidence based on limited data that host susceptibility in natural habitats interacts with plant type; exophytic and leaf-mining hosts become even more susceptible as succession proceeds in natural habitats, whereas less susceptible hosts suffer reduced parasitoid-induced mortality in late succession. No such interactions are apparent in cultivated habitats.

There are no latitudinal gradients in host mortality levels. Tropical herbivores suffer as much parasitism as temperate herbivores, even when the number of parasitoid species is lower. This lends further support to the hypothesis that most of the parasitoids in speciose systems contribute little to host mortality, in this case suggesting that the additional species occurring in temperate host–parasitoid systems are dynamically unimportant generalists. It also follows that parasitoids in general are equally important mortality factors in all parts of the world and that parasitoids are fully capable of finding hosts that may inhabit fragmented, tropical habitats.

Biological control success rates also vary among host feeding niches, and the form of the relationship is similar to that found for parasitoid species richness and host mortality. Consequently, all three host–parasitoid variables are intercorrelated; types of herbivore that generally support more parasitoid species and suffer higher maximum parasitoid-induced mortality are also more frequently controlled following parasitoid introductions. The susceptibility hypothesis, which argues that refuges from parasitism represent the dominant constraint on host–parasitoid interactions, provides a parsimonious explanation for all three patterns.

6
Hyperparasitoids

6.1 Introduction

Up to this point I have dealt exclusively with the patterns found for primary parasitoid complexes and their herbivorous hosts. However, parasitoids themselves are often subject to attack by obligatory hyperparasitoids (as well as facultative hyperparasitoids), and these hyperparasitoid complexes can also be quite rich, sometimes rivaling the species richness of the primary parasitoids. For example, the teak defoliator *Pyrausta machaeralis* (Walker) (Lepidoptera: Pyralidae) has 42 reported species of obligatory hyperparasitoids and 42 primary parasitoids in Burma (Garthwaite & Desai, 1939). Occasionally, hyperparasitoid species richness may even greatly exceed that of the primaries. The citrus psylla, *Trioza erytreae* (Del Guercio) (Homoptera: Psyllidae), supports two primary parasitoids and 13 species of hyperparasitoids and tertiary hyperparasitoids (i.e. parasitoids of the hyperparasitoids) in South Africa (McDaniel & Moran, 1972). But despite these few examples, it appears that in general hyperparasitoid richness is lower than primary richness, and many parasitoid complexes may be entirely free of obligatory hyperparasitoids.

A major problem when attempting to evaluate hyperparasitoid complexes is that they have been relatively poorly studied, with most of the emphasis on working out complex, sometimes bizarre biologies of individual species. The hyperparasitoids of Homoptera have received the most attention (see review in Sullivan, 1987), whereas those of holometabolous herbivores have been more erratically studied. Ecological studies of non-homopteran hyperparasitoid communities are relatively uncommon and, because of ease of study, usually deal more with pseudohyperparasitoids than with true hyperparasitoids (Muesebeck & Dohanian, 1927; Pike & Burkhardt, 1974; Sickle & Weseloh, 1974; Morris, 1976; Weseloh, 1979, 1986; Elliott, Simmons & Haynes, 1986; Bourchier & Nealis, 1992). Much of the discussion of hyperparasitoids has focused on their potential interference with herbivore population regulation and biological control (Narayanan & Subba Rao, 1960; Flanders, 1963; Rosen, 1981; Weseloh, 1983; Agricola & Fischer, 1991), and in this context they have received theoretical treatment in models of host–parasitoid population dynamics (Nicholson & Bailey, 1935; Beddington & Hammond, 1977; Hassell, 1978; Luck, Messenger & Barbieri, 1981; May & Hassell, 1981).

Given that relatively few studies have been explicitly designed to study hyperparasitoid communities, how accurate are estimates of hyperparasitoid species richness likely to be? Unless a hyperparasitoid species belongs to a higher taxon known or at least believed to be universally hyperparasitic (e.g. mesochorine Ichneumonidae, Shaw 1993), many will not even be recognized as such unless parasitoid rearings are conducted with considerable care and control. Further, many species may be rare relative to the herbivore or their primary parasitoid hosts and will escape detection entirely in less intensive studies. Therefore, hyperparasitoid species richness will be severely underestimated in many cases. On the other hand, many idiobionts are facultatively hyperparasitic; if only one to a few individuals are encountered, they may be mistakenly classified as obligatory hyperparasitoids when they are not. In these cases, hyperparasitoid species richness will be overestimated. For all of the above reasons, it is inevitable that data on hyperparasitoid species richness will be much more variable in quality and more error prone than those for primary parasitoids.

With these potential problems with the data in mind, can we expect to be able to identify patterns of hyperparasitoid species richness? In this chapter I explore patterns in the fourth trophic level as an extension of the analyses of primary parasitoids. The independent variables are those used in the analyses of the primary parasitoids, and the approach is similar, with some exceptions. First, the absence of obligatory hyperparasitoids reported in a system may reflect biological reality, or it may mean that no attempts were made to identify them (or authors simply did not report them); that is, a '0' may be real or an artefact. Because some 0s undoubtedly are real whereas others undoubtedly are not, all analyzes were done twice, first using all herbivores (i.e. assuming that all 0s are real, $n = 2188$), and second using only those systems in which at least one hyperparasitoid was reported (i.e. assuming that all 0s are artefacts of incomplete study, $n = 328$). This brackets the possibilities, and if similar results are found using both approaches, they are probably robust. But if results differ qualitatively, it is difficult to judge which data set is most reliable, and interpretations will have to be tentative until more data are available.

The measure of hyperparasitoid species richness used in most analyses is the number of species per herbivore species, because true hyperparasitoids attack the primary parasitoids through the herbivore, and characteristics of the herbivore would be expected to have the most effect on hyperparasitoid species richness patterns. Most of the data come from studies based on the rearing of herbivores, with fewer rearings of field collected parasitoid cocoons that may have been subject to attack by pseudohyperparasitoids.

Finally, the distribution of hyperparasitoid species richness was extremely skewed in both the data excluding 0s and those including 0s, which was not corrected by log-transformation. Thus, non-parametric statistics are used for the bulk of analyses that used the number of hyperparasitoid species as the dependent variable. Nevertheless, all means reported are geometric, irrespective of whether parametric or non-parametric statistics were used.

6.2 Primary parasitoid species richness

The simplest pattern of hyperparasitoid species richness that should exist is a relationship between the number of hyperparasitoid species in a system and the number of primary parasitoid species. This could arise for at least three reasons: (a) more primary species represent more potential resources for hyperparasitoids; (b) hyperparasitoids search for herbivores that support their primary-parasitoid hosts in much the same way as do the primary parasitoids, so that herbivores likely to be found by many primary parasitoids will also be found by more hyperparasitoids; and (c) more intensive studies will detect more of both types of parasitoid. So the first question is whether the presence of hyperparasitoids is associated with the number of primaries. Dividing the complete global data into those systems with no hyperparasitoids and those with at least one species, the systems with hyperparasitoids are more than twice as rich in primaries as those with no hyperparasitoids (geometric means = 7.48 and 3.27, respectively; ANOVA, $F = 266.76$, $P < 0.001$). But it is possible that this is a sampling artefact, since increasing sample size should increase numbers of species of both types of parasitoid, and hyperparasitoids may have been overlooked entirely in poorly sampled systems. This possibility was examined with an ANCOVA of the 452 cases where sample sizes are known. After a preliminary test of the homogeneity of slopes ($P = 0.786$), the number of primaries still had a highly significant relationship with the presence of hyperparasitoids after adjusting for sample size as a covariate ($F = 66.20$, $P < 0.001$; adjusted means: hyperparasitoids absent = 4.23 primaries, n = 346; hyperparasitoids present = 8.70 primaries, $n = 105$). Sampling intensity appears to have a minimal influence on the result.

Given that the presence of hyperparasitoids is associated with the number of primaries, is there a relationship between the number of species of hyperparasitoids and the number of species of primaries? Indeed, there is a noisy, but

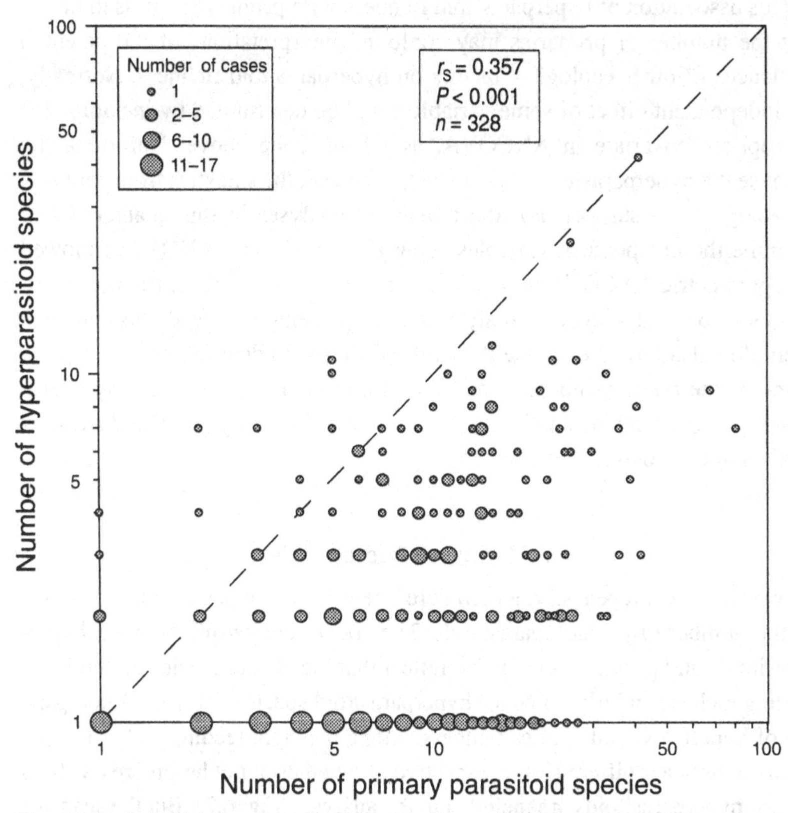

Fig. 6.1. Relationship between hyperparasitoid species richness and primary parasitoid species richness for 328 parasitoid complexes. Dashed line is line of unity.

highly significant association as indicated by Spearman rank correlation (Fig. 6.1). To test for the possible influence of sampling intensity, it is necessary to use a parametric technique (multiple regression) on data that are badly skewed. But Pearson's product–moment correlation gives a very similar result to that obtained using rank correlation ($r_P = 0.377$ versus $r_S = 0.357$) suggesting that violating the assumption of normality does not strongly alter the relationship based on such a large number of cases. Using the residuals from the regression of the number of hypers (log-transformed) on sample size (log-transformed) (which are associated, $r = 0.331$, $F = 12.660$, $P < 0.001$, $n = 105$), the relationship between the numbers of hyperparasitoids and primaries remains positive and significant ($r = 0.280$, $F = 8.79$, $P = 0.004$). Therefore, the first pattern is that the presence and number of species of hyperparasitoids in a herbivore–parasitoid system is positively associated with the number of primary parasitoids, independently of how intensively the system is sampled.

This association of hyperparasitoid richness with primary richness indicates that the number of primaries may confound interpretations of the potential influences of other ecological factors on hyperparasitoid richness. Normally, the independent effect of some variable could be determined by including the appropriate covariate in ANCOVA, as I have done above. Unfortunately, because the hyperparasitoid data are badly skewed, the statistics from parametric analyses are suspect. For the remaining analyses in this chapter, I first examine the independent variables using Kruskal–Wallis ANOVA, followed by a parametric ANCOVA with primary richness as a covariate. But the probabilities associated with these analyses are not presented. Instead, this approach is employed solely to examine the adjusted means to determine if the general shape of the relationship between each independent variable and hyperparasitoid richness remains qualitatively similar when primary parasitoid richness is taken into account.

6.3 Herbivore feeding niche

As we have seen repeatedly, a herbivore's feeding niche has a major influence on the number of primary parasitoids. The above relationship between hyperparasitoids and primaries could also reflect that the characteristics of herbivore feeding niches similarly influence hyperparasitoid species richness. A comparison of hyperparasitoid species richness among the eight feeding niches reveals highly significant differences, irrespective of whether or not herbivores with no known hyperparasitoids are included in the analysis (Fig. 6.2). But the shape of the relationship does not simply mirror the dome-shaped pattern found for primaries. Instead, hyperparasitoid species richness falls into two fairly distinct groups, with hyperparasitoids being relatively rich in four herbivore niches and being relatively poor in the other four (Fig. 6.2). Further, the species-rich assemblages occur in all of the niches in which the herbivores are entirely or partially exophytic, whereas the hyperparasitoid-poor systems are found in those niches wherein herbivores are either endophytic or otherwise concealed (the root feeders). It is notable that leaf miners, which support among the richest primary parasitoid complexes, support few obligatory hyperparasitoids.

Using ANCOVA to examine the pattern when primary species richness is accounted for, the pattern for all herbivores remains qualitatively similar, although the quantitative differences between the two basic types of host are smaller (Fig. 6.3*a*). Completely or partially exophytic herbivores support relatively rich hyperparasitoid assemblages and endophytic/root-feeding herbivores support relatively depauperate ones. When comparing only those systems supporting hyperparasitoids, the pattern is also similar, except for the root

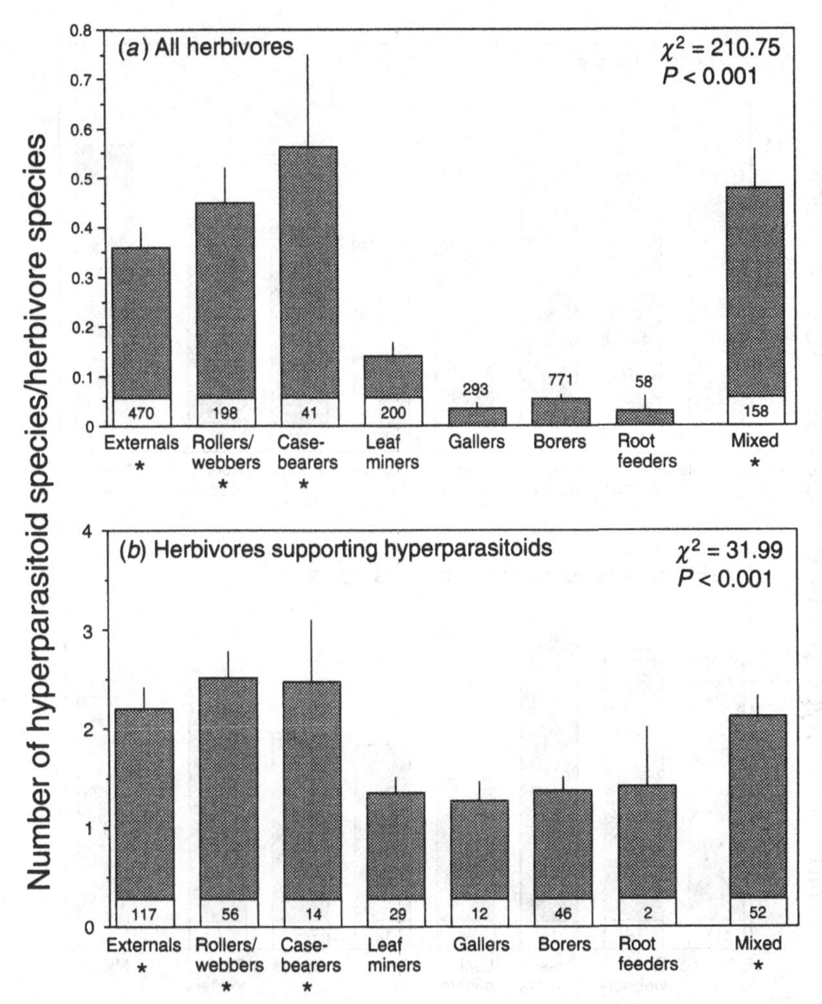

Fig. 6.2. Relationship between host feeding niche and mean hyperparasitoid species richness for (*a*) all herbivore species and (*b*) only those herbivores for which at least one hyperparasitoid has been reported. Numbers at the base of the bars are the number of host species in each niche; vertical lines are +1 S.E.M. The first seven niches are ranked in terms of decreasing mobility and increasing concealment in foodplant or soil. Asterisks indicate those niches comprising completely or partially exophytically feeding herbivores.

feeders which seem to support relatively many species. But this represents a distortion due to a small sample size (two herbivore species with hyperparasitoids; 3.4% of all root feeders) coupled with a single 'hyper-rich' system. *Cleonus mendicus* (Gyllenhal) (Coleoptera: Curculionidae) is reported to support two hyperparasitoids on a single larval primary (Isart, 1972). This strongly

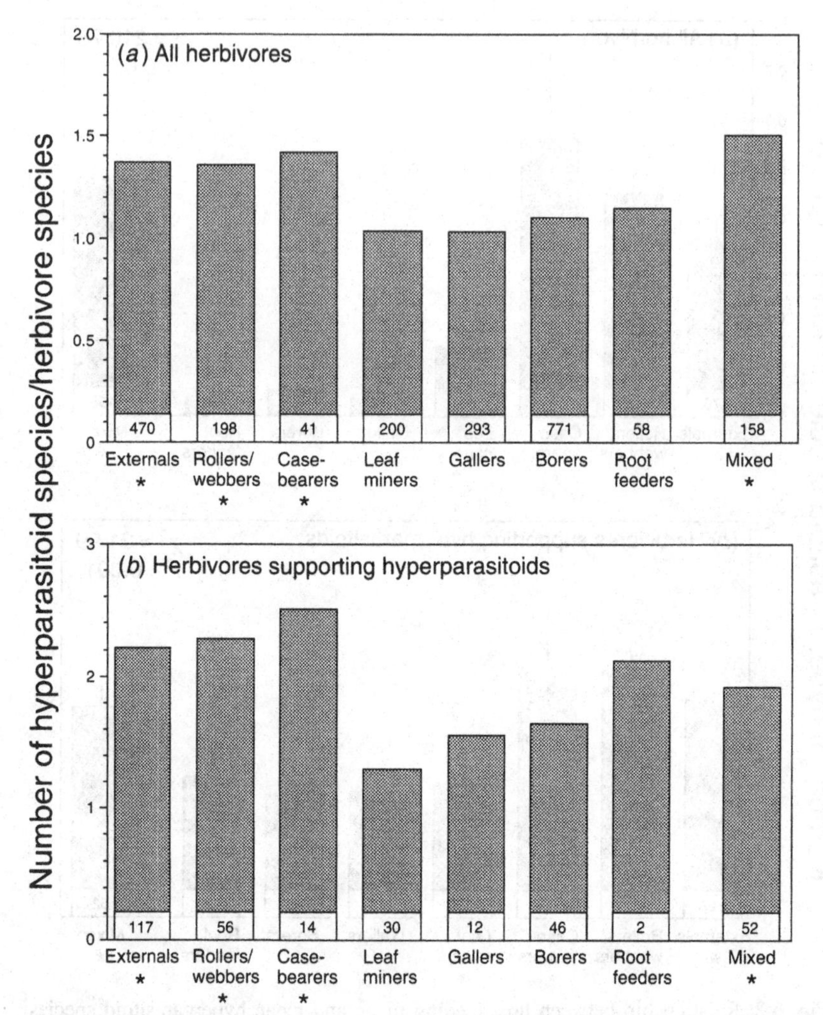

Fig. 6.3. Hyperparasitoid species richness after adjusting for the number of primary parasitoid species for (*a*) all parasitoid complexes and (*b*) those complexes supporting at least one hyperparasitoid.

influences the adjusted mean, making root feeders appear to support rich hyperparasitoid complexes relative to their primaries, when they are actually rarely reported at all (see Fig. 6.2a). Despite this case, the analyses in general indicate that the second pattern in hyperparasitoid species richness represents an influence of herbivore feeding niche. But the relationship is much simpler than that for primaries, depending largely on whether or not the herbivore through which most of the hyperparasitoids are locating their primary parasitoid hosts is exophytic (or otherwise exposed) for at least part of this larval development or is completely endophytic (or otherwise concealed).

6.4 Foodplant/habitat

Primary parasitoids show complex interactions with the type of plant and the type of habitat in which their hosts occur. With the possible exception of gallers and borers, primary complexes tend to be richer on hosts on trees than on hosts on herbs in natural habitats, whereas in cultivated habitats plant type effects disappear. Further, concentrating resources by cultivating herbs increases parasitoid species richness relative to herbs growing naturally, whereas reducing plant diversity by cultivating trees reduces parasitoid richness relative to trees growing naturally (Section 3.3). Do hyperparasitoid complexes respond to plant type and habitat similarly?

Because the feeding niche pattern for hyperparasitoids indicates that they distinguish only two basic types of herbivore, plant/habitat effects are analyzed in terms of 'exposed' herbivores (i.e. exophytics) and 'concealed' herbivores (i.e. endophytics and root feeders). This approach also increases sample sizes, decreasing the importance of idiosyncratic patterns bound to arise from dividing the data into small subsets (as occurred in the preceding analysis).

When all herbivores are included, absolute hyperparasitoid richness (the number of species per herbivore species) is highest on exposed herbivores on trees in natural habitats, whereas in cultivated habitats exposed herbivores on monocots support the most species (Fig. 6.4*a*,*b*). But when the number of primaries is accounted for, there are no strong patterns in natural habitats (Fig. 6.4*c*). This suggests that in natural communities, hyperparasitoid richness simply follows primary richness and there are few independent effects of plant type. In cultivated habitats, on the other hand, hyperparasitoid richness remains high on monocots relative to the number of primaries (Fig. 6.4*d*). Hyperparasitoid richness varies little on concealed hosts, being very low on all plant types in both natural and cultivated habitats.

Considering only those systems that are reported to support hyperparasitoids, absolute richness shows a very similar pattern (Fig. 6.5), with richness greatest on exposed herbivores on trees in natural habitats and greatest on exposed hosts on monocots in cultivated habitats. Accounting for primary richness does not alter the patterns substantially, although the difference between trees and non-trees is smaller in natural habitats. Therefore, the third pattern in hyperparasitoid species richness, if it can be called that, is that independent effects of plant type are present, but weak, and hyperparasitoid richness patterns on each type of plant for the most part mirror those of their primary parasitoid hosts.

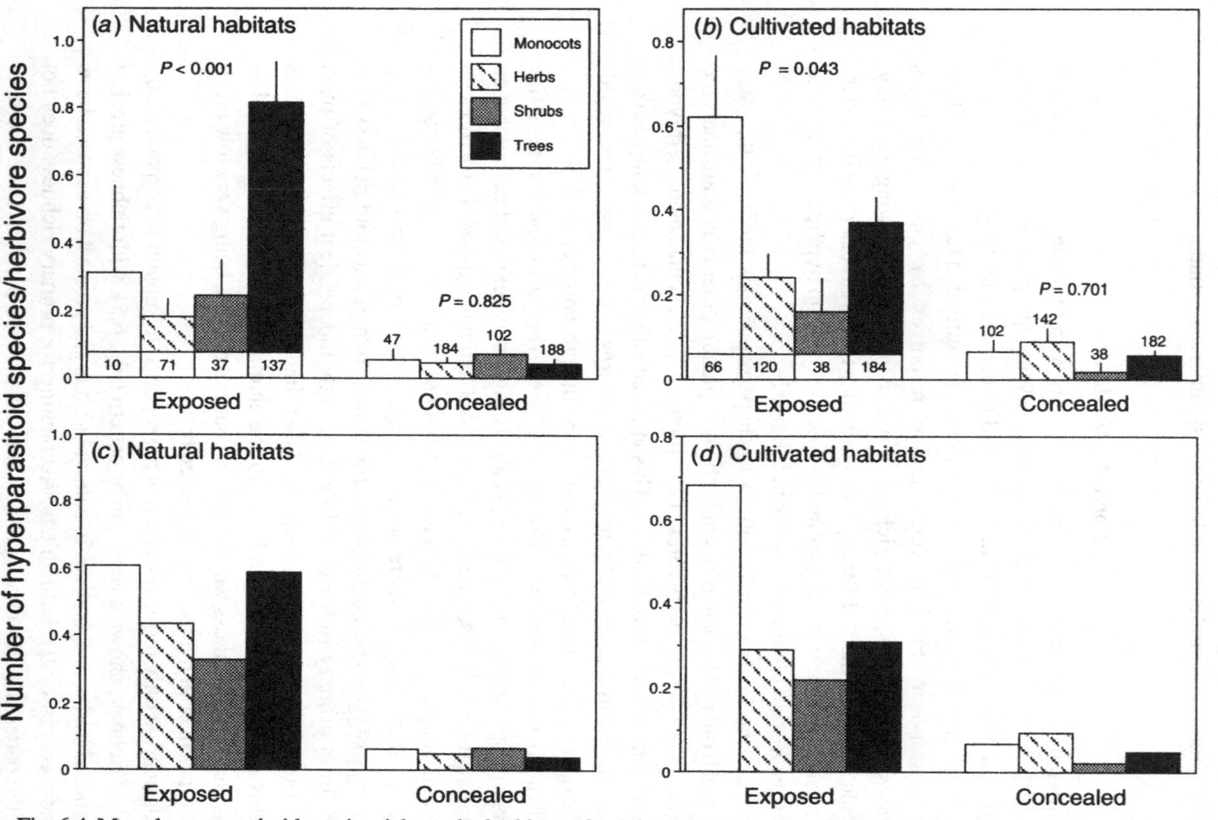

Fig. 6.4. Mean hyperparasitoid species richness by herbivore-foodplant type in (*a*, *c*) natural and (*b*, *d*) cultivated habitats, including all herbivores. (*a*, *b*) Raw number of hyperparasitoid species per herbivore species and (*c*, *d*) number of hyperparasitoids per herbivore species adjusted by the number of primary parasitoids. Herbivores are distinguished as being either exposed (external folivores, rollers/webbers, casebearers and mixed exo-/endophytic feeders) or concealed (leaf miners, gallers, borers and root feeders).

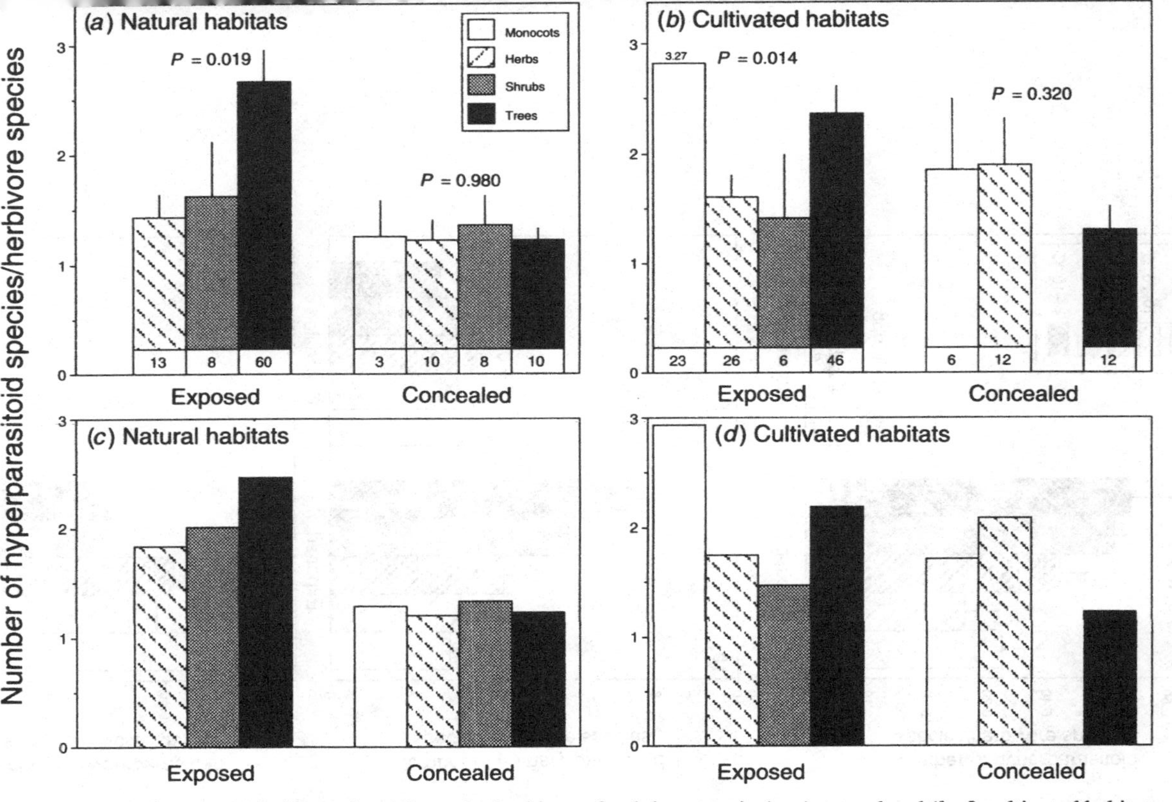

Fig. 6.5. Mean hyperparasitoid species richness by herbivore-foodplant type in (*a*, *c*) natural and (*b*, *d*) cultivated habitats, including only those herbivores supporting hyperparasitoids. (*a*, *b*) Raw number of hyperparasitoid species per herbivore species and (*c*, *d*) number of hyperparasitoids per herbivore species adjusted by the number of primary parasitoids. Herbivores distinguished as in Fig. 6.4.

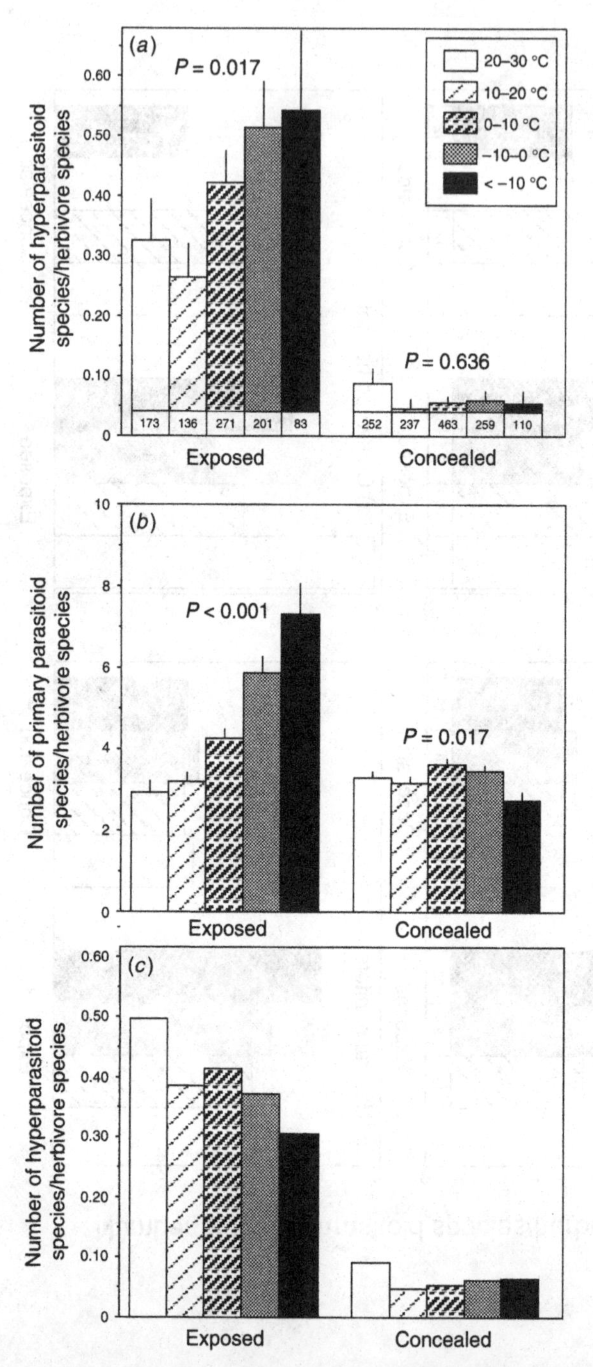

Fig. 6.6. Climatically-based gradients in hyperparasitoid and primary parasitoid species richness, using mean temperature in the coldest month, for all exposed and concealed herbivores. (*a*) Number of hyperparasitoid species, (*b*) number of primary parasitoid species, and (*c*) number of hyperparasitoid species adjusted by the number of primary parasitoids.

6.5 Latitude/climate

We have seen that among primary parasitoids, species richness tends to decrease towards the tropics on hosts that are completely or partially exophytic, whereas it does not for endophytic/root-feeding hosts. To examine latitudinal variation in hyperparasitoid species richness, I use mean temperature in the coldest month as the climatic variable and again distinguish hosts as being either exposed or concealed.

First, analyzing all herbivores indicates that absolute hyperparasitoid species richness per herbivore species also decreases towards the tropics on exposed hosts, or at least that species richness is highest in the northern temperate zone (Fig. 6.6*a*). Species richness does not vary significantly among climatic zones on concealed hosts. Of course, primary parasitoids also show different patterns on the two types of host, falling on exposed hosts and remaining fairly constant on concealed hosts (Fig. 6.6*b*), as would be expected from the analyses of the eight individual feeding niches. When hyperparasitoid species richness is adjusted for primary richness by ANCOVA, it actually *increases* towards the tropics on exposed hosts (Fig. 6.6*c*)! It is also highest on tropical concealed hosts, although only marginally so. Considering only those systems supporting hyperparasitoids, there are no significant climatic gradients, although assemblages on exposed herbivores tend to be richest in the tropics in absolute terms (Fig. 6.7*a*). The richness of hyperparasitoids in the tropics becomes pronounced when the number of primaries is taken into account (Fig. 6.7*c*). As in all previous analyzes, there are no strong relationships on concealed hosts, although tropical richness tends to be slightly higher in both absolute (Figs. 6.6*a*, 6.7*a*) and relative (Figs. 6.6*c*, 6.7*c*) terms.

These analyses suggest that the absolute number of hyperparasitoids may or may not vary across a climatic gradient, depending on whether the data set including or excluding 0s more accurately reflects reality. However, in relation to the primary parasitoids that serve as their hosts, hyperparasitoid assemblages are as rich or perhaps even richer in the tropics than in the non-tropics, particularly in exposed herbivores where most species occur. The fourth pattern in hyperparasitoid species richness is that they are proportionately better represented in tropical systems than in temperate ones.

6.6 Conclusions

Perhaps the most surprising result arising from the analysis of hyperparasitoids is that there are any discernible patterns at all. Despite the highly variable quality of the data, reasonably clear relationships exist between hyperparasitoid

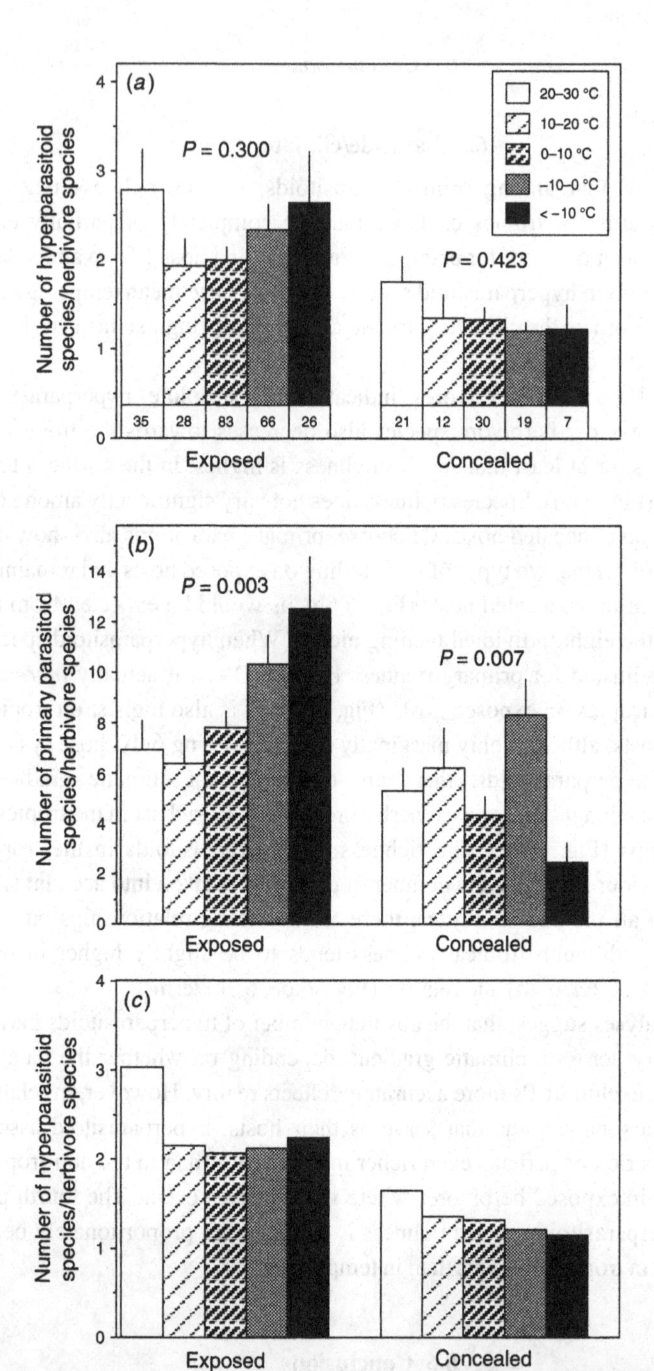

Fig. 6.7. Climatically-based gradients in hyperparasitoid and primary parasitoid species richness, using mean temperature in the coldest month, for those exposed and concealed herbivores that support hyperparasitoids. (*a*) Number of hyperparasitoid species, (*b*) number of primary parasitoid species, and (*c*) number of hyperparasitoid species adjusted by the number of primary parasitoids.

species richness and biological and ecological characteristics of the herbivores. Further, when measures of sampling intensity are incorporated into some of the analyses, the patterns are qualitatively robust. The data may not be as bad as imagined and, although prone to error, are good enough to detect underlying relationships.

A positive relationship between the numbers of hyperparasitoid species and primary parasitoids is not particularly surprising. If for no other reason, the more intensively a system is studied the more likely hyperparasitoids will be encountered and recognized as such. Similarly, better study will reveal more primary parasitoid species. This may be true, but the observed relationship does not appear to reflect solely the sampling intensity. The relationship persists after incorporating sample size. Another possible explanation is that larger numbers of primary parasitoids provide additional potential hosts for hyperparasitoids, elevating the latter's diversity. However, the pattern among herbivore feeding niches suggests that this explanation is insufficient. Leaf miners support many primary parasitoids, but few obligatory hyperparasitoids, so primary parasitoid species richness *per se* is not that important. The most viable explanation is that the biological and ecological characteristics of hosts that have been shown to constrain primary parasitoids similarly constrain hyperparasitoids, and the extent that herbivores and/or primary parasitoids are exposed to attack similarly influences the fourth trophic level. For example, Shaw & Aeschlimann (1994) have found that the mesochorine *Mesochorus ?crassicrus* Thomson can develop in a large number of koinobiont primary parasitoid species, but its lepidopteran 'host' range is restricted to caterpillars feeding under webs. Therefore, in this case herbivore feeding behavior represents the major constraint on the hyperparasitoid, which, once it has located the caterpillar, will simply attack virtually any primary parasitoid present. So we are lead to the conclusion that, as with primary parasitoids, hyperparasitoids are strongly influenced by herbivore feeding biology. However, in the case of hyperparasitoids, the relationship between species richness and feeding niche is much more straightforward than that found for primary parasitoids, and simply feeding by herbivores within plant tissues or hidden in the soil appears sufficient to limit them. There are several factors that might contribute to this pattern.

First and foremost, much of the difference in hyperparasitoid species richness between exophytic and endophytic herbivores is due to the biologies of the species involved and the way that the parasitoid data were scored. The primary parasitoids of exophytic herbivores represent mainly koinobionts attacking larvae (Chapter 4). Consequently, the hyperparasitoid complexes of these primaries will comprise both koinobiont true hyperparasitoid attacking primary parasitoid larvae within herbivore larvae and idiobiont pseudohyperparasitoids attacking parasitoid cocoons that have formed within herbivore cadavers or in their vicinity. The combination of both types of hyperparasitoids results in rich complexes. The primaries of endophytic herbivores, on the other hand, comprise mainly idiobionts. At least partially because of their concealment (see below), these are likely to support relatively few koinobiont hyperparasitoids, which leaves just the idiobiont component. Many (most?) of these idiobionts are facultatively hyperparasitic, which I do not distinguish from purely primary parasitoids, and have not been included in the hyperparasitoid data. Actually, hyperparasitism is rampant in endophytic host–parasitoid systems, but most of it represents facultative hyperparasitism by idiobionts. This is in contrast to exophytic systems, which in addition to pseudohyperparasitoids also support many more koinobiont true hyperparasitoids.

As alluded to above, part of the difference in hyperparasitoid species richness in exophytic and endophytic systems probably reflects real effects of host concealment. For many (but not all) koinobiont obligatory hyperparasitoids, direct contact with the primary parasitoid larva may be necessary for successful oviposition, and anything that interferes with this contact disrupts them. Even the herbivore's cuticle may represent a partial barrier to those hyperparasitoids that must locate a larva of its primary-parasitoid host within the herbivore's body cavity to oviposit into it. The addition of a second barrier to oviposition, even when it is only leaf epidermis, may simply be too difficult for many hyperparasitoids to overcome.

At least some hyperparasitoids are extremely sensitive to physical barriers to oviposition, although I am not aware of any studies that have examined the influences of plant structures. The aphid hyperparasitoid *Alloxysta victrix* (Westwood) must attack its primary-parasitoid host *Aphidius smithi* Haliday when the aphid is alive, because once the latter has mummified the parasitoid larva is 'inaccessible' (Gutierrez & van den Bosch, 1970); *Tachinobia repanda* Boucek apparently attack pupae of their tachinid host only when either the lepidopteran pupal case (in which the tachinids have pupated already) has a hole in it or when the tachinid kills the caterpillar before it has had a chance to form a cocoon (Mayer & Shull, 1978); and *Cheiloneurus noxius* Compere is sometimes unable to attack its host *Metaphycus lounsburyi* (How.) when the scale covers through which females must drill 'prove too tough' (Le Pelley, 1937). I have argued throughout this monograph that host inaccessibility, partially due to the protection provided by plant parts, represents a strong constraint on primary parasitoid species richness and host utilization rates. It is likely that hyperparasitoids are equally affected.

It is perhaps not surprising that, other than the plant-derived effect outlined above, plant/habitat effects appear to have minimal impacts on hyperparasitoids. Ecological and biological constraints arising from the first trophic level should be strongest on the second trophic level (i.e. on the herbivores), but they should become progressively weaker as the number of links between resource and consumer increases. Relationships of the primary parasitoids with plant and habitat type already appear somewhat weak and idiosyncratic (Section 3.3), and their influences have probably been largely dissipated by the time they reach the fourth trophic level. The few patterns that are apparent suggest that both primary parasitoids and hyperparasitoids are influenced by plant type and habitat type in much the same way, and that no special mechanisms unique to hyperparasitoids appear to be operating.

The most striking pattern is the latitudinal/climatic gradient. Relative to the primary parasitoids, tropical communities on average support more hyperparasitoids than temperate communities, irrespective of the data set analyzed. This raises two questions, the first relating to the overall quality of tropical versus temperate zone parasitoid data, and the second relating to the mechanistic basis for latitudinal gradients in parasitoid diversity.

Despite attempting to control for the degree of knowledge of parasitoid complexes in all parts of the world and checking the robustness of the latitudinal gradients using subsets of data, can we rule out completely the possibility that the loss in species richness found among the parasitoid complexes associated with exophytic hosts reported in Chapters 3 and 4 somehow reflects the fact that temperate communities are simply better known than tropical communities? The hyperparasitoid data argue against this possibility. If the climatic patterns actually represent a gradient in knowledge rather than biological reality, we would expect hyperparasitoids to show even stronger gradients. If the data on primary parasitoid complexes are to be considered poor and latitudinally biased, then those for hyperparasitoids must be even worse, since even more basic understanding of the biologies of the species involved and more detailed rearing techniques are necessary to recognize that a parasitoid is acting as a hyperparasitoid. It seems very difficult to imagine that tropical workers could underestimate the richness of the primary parasitoid community through incomplete study and yet report higher proportions of hyperparasitoids. One conclusion seems unescapable: if the latitudinal data are not biased, hyperparasitoid assemblages are somewhat richer in the tropics; if the data are biased, and further work reveals many additional primary parasitoids in tropical systems, it is likely that the latitudinal gradient for hyperparasitoids will be found to be even stronger. Therefore, irrespective of whether the quality of the tropical data is low relative to the temperate data, it appears that tropical communities are rich in hyperparasitoids.

The second question raised is which, if any, of the potential mechanisms that hope to account for latitudinal gradients in parasitoid species richness is most consistent with the hyperparasitoid pattern? Are the current hypotheses already examined to explain primary parasitoid species richness (Chapter 4) sufficient to also explain hyperparasitoid richness, or to turn the question around, does the hyperparasitoid pattern lend additional support to one over another? The mechanisms most likely to be relevant are resource fragmentation and nasty hosts; the first because it assumes that many tropical parasitoids are unable to locate enough hosts to maintain their populations, and the second because primary parasitoids feeding on nasty hosts could themselves sequester noxious plant allelochemicals, which could then affect the fourth trophic level.

If resource fragmentation is strong enough to limit the number of tropical species of primary parasitoids, it would be expected to be even more pervasive for hyperparasitoids, all else being equal. Hyperparasitoids, particularly true hyperparasitoids, should be under severe dynamic constraints, since they must not only find the appropriate herbivores, but must locate herbivores that also have been, or will be, found by a primary parasitoid. Even pseudohyperparasitoids should be affected, since the population of parasitized herbivores will be smaller than the total herbivore population. If resource fragmentation is critical in the tropics, hyperparasitoid diversity might thus be expected to be extremely low. Whether or not absolute hyperparasitoid species richness is lower in the tropics depends on which data set is examined (*cf.* Figs. 6.6*a* and 6.7*a*). However, there appears little doubt that relative to the primary parasitoids (Figs. 6.6*b*, 6.7*b*), any reduction in hyperparasitoid species richness is less severe than that for the primaries, and once the number of primaries has been factored out, hyperparasitoid species richness is actually greatest in the tropics (Figs. 6.6*c*, 6.7*c*). On the face of it, this is clearly inconsistent with the resource fragmentation hypothesis.

A critical piece of missing information that is necessary for evaluating resource fragmentation is the relative host ranges of primary parasitoids and hyperparasitoids. It could be argued that hyperparasitoids might be able to offset the effects of resource fragmentation if they are able to expand their host ranges sufficiently in the tropics. Until relatively recently, it was believed that hyperparasitoids are broadly polyphagous (Gutierrez & van den Bosch, 1970; Gordh, 1981), and many undoubtedly are. It remains possible that hyperparasitoids are less affected by resource fragmentation than primary parasitoids and are thus able to maintain a relatively high tropical diversity. On the other hand, it has become clear that some hyperparasitoids are quite specific (Gutierrez & van den Bosch, 1970; van den Bosch, 1981). If hyperparasitoids show a similar distribution of host ranges as do primary parasitoids, then resource fragmentation cannot account for the former's latitudinal gradient; but if most hyperparasitoids are in fact broadly polyphagous, they may be relatively immune to resource fragmentation. It is important to realize that for the latter to be true, they must be polyphagous at both 'host' levels: they must attack a wider range of herbivores than primary parasitoids *and* a wide range of primary parasitoids associated with the herbivores. As is the case for all aspects of parasitoid community structure, until better host range data are available it is difficult to judge the importance of presumed underlying mechanisms.

As mentioned above, the nasty host hypothesis has ramifications for hyperparasitoid as well as primary parasitoid diversity patterns. There are two possibilities. First, if primary parasitoids also sequester the noxious chemicals found in their hosts, hyperparasitoid species richness should be low in the tropics relative to the number of primaries. Of course, this is not what is found, so if nasty hosts are much more frequent in the tropics, as argued by the hypothesis, this parasitoid sequestration must not be common. Second, if primary parasitoids utilizing nasty hosts often detoxify plant-derived allelochemicals (as the parasitoids of the cyanide-containing caterpillars of *Zygaena filipendulae* (L.) are capable of doing (Jones, Parsons & Rothschild, 1962)), they could act as 'filters', releasing hyperparasitoids from the constraints that they themselves are under. Again, if resource fragmentation is not too strong or if hyperparasitoids are able to offset its effects by expanding their host ranges, hyperparasitoids are then free to proliferate in the tropics.

Due to a lack of data, it is not possible to unequivocally determine if resource fragmentation, nasty hosts, or a combination of both are best supported by hyperparasitoid species richness patterns. However, I suspect that plant chemistry may be at least as important as resource fragmentation based on the sum of the patterns found for both primary parasitoids and hyperparasitoids. Needless to say, additional data on the ecologies and host ranges of hyperparasitoids would be helpful.

6.7 Summary

The species richness of obligatory hyperparasitoids is typically lower than that of the primary parasitoids that compose a complex, although the diversity of the former is likely to be poorly documented unless parasitoid rearings are done carefully. However, despite the problems with the data arising from the erratic study of hyperparasitoid complexes, it does appear possible to identify at least some of the constraints on their species richness.

Foremost, the number of hyperparasitoid species is positively associated with the number of primary parasitoids, a relationship that is not simply due to variability in intensity of study. Consequently, many of the factors that influence primary parasitoid species richness show similar relationships with hyperparasitoids.

Exophytic herbivores generally support more hyperparasitoids than endophytics, probably reflecting a paucity of koinobiont true hyperparasitoids on endophytics due to a combination of greater numbers of idiobiont facultative hyperparasitoids and barriers to oviposition arising from host concealment. Hyperparasitoids respond to plant and habitat types, but the relationships are weak and for the most part disappear when the number of primary parasitoids is factored out.

It is unclear whether or not there is a latitudinal gradient in the absolute number of hyperparasitoids. But relative to the number of primary parasitoids, the number of hyperparasitoid species is higher in the tropics than in the extratropics. This cannot arise from a latitudinal gradient in the quality of the data unless tropical workers have focused more on hyperparasitoids than their temperate counterparts, which seems unlikely. The mechanism(s) underlying this gradient is(are) difficult to evaluate because hyperparasitoid host ranges are poorly documented, but hyperparasitoids may be less affected by resource fragmentation than are primary parasitoids, and primary parasitoids may not sequester the plant-derived allelochemicals found in tropical folivores, releasing their hyperparasitoids from this constraint.

7
Synthesis

The major goal of this work has been to determine if it is possible, through the haze of imperfect data and a mass of ecological and evolutionary complexity, to identify general patterns in host–parasitoid interactions. My approach is in stark contrast to that of others who have rightly pointed out the complex and idiosyncratic nature of relationships involving a majority of the described species on the planet (see for example, Price, 1994). Undoubtedly, the weakness of some of the patterns found does reflect the fact that ecological communities are complex, and that much more focused studies are required for discerning underlying relationships in particular host–parasitoid groups. On the other hand, it does not follow that the forces structuring communities are so varied and convoluted that attempts to derive general principles are hopeless. Reasonably clear and consistent relationships between a herbivore's feeding niche and the size and composition of its parasitoid complex, the maximum parasitism rate it suffers, the probability that parasitoids will severely depress its density, and even the species richness of its hyperparasitoids, identify factors related to host feeding biology as being of fundamental importance to the interactions between parasitoids and their hosts. If the patterns found in my data are as robust as they appear, I believe that feeding niche can be used as a template on which to build a general theory of host–parasitoid interactions. This is not to say that other ecological and evolutionary factors are not also operating, but the available data suggest that these additional forces are secondary and operate independently of the feeding niche template.

Figure 7.1 illustrates what a 'feeding niche' model of host–parasitoid interactions might look like, based on the patterns known to date. The host's feeding biology constrains the number of parasitoids that will utilize that host (Section 3.2) and is similarly associated with the amount of mortality parasitoids are able to inflict (Section 5.2). Importantly, the association between parasitoid species richness and host mortality is indirect, acting through the

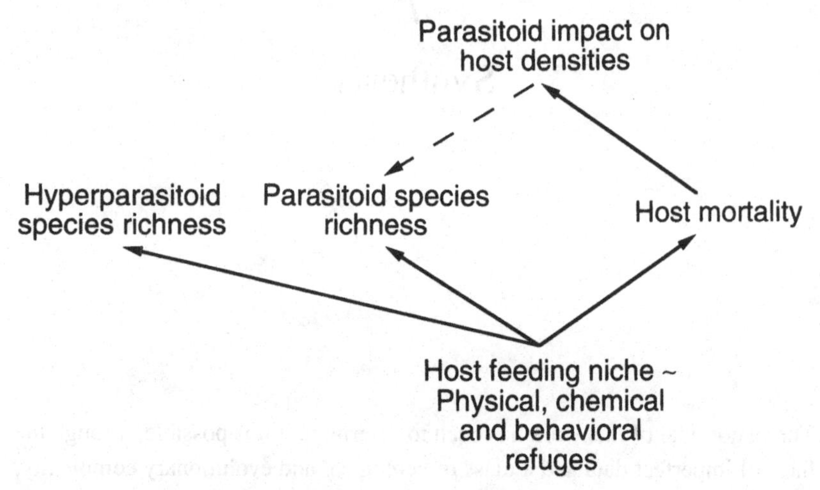

Fig. 7.1. A simple model of host–parasitoid–hyperparasitoid interactions centered on herbivores' feeding niches, which are assumed to represent a proxy variable for the extent of host refuges from parasitoid attack.

inherent differences in average susceptibility of hosts in each feeding niche, rather than being directly linked (Section 5.6). The association between host mortality and the impact of parasitoids on host densities (Section 5.5), on the other hand, is presumed to be direct, since hosts that are attacked by parasitoids at very low levels are unlikely to suffer much density depression due to the latter's actions. Although evidence for a relationship between host density and parasitoid species richness has to be more tentative since it is based on the analysis of sample size rather than on direct measures of abundance (Section 3.2), parasitoid impact would be expected to feed back on parasitoid species richness: hosts that are kept scarce by parasitoids would support fewer species. The parasitoids that are lost would likely be opportunistic generalists that are not adapted to searching for specific rare host species and would encounter fewer individuals of that herbivore species in the pool of potential host species.

Hyperparasitoid communities can also be incorporated into this model. The number of hyperparasitoid species is positively associated with the number of primary parasitoids (Section 6.2), but underpinning this association is whether the herbivore is exophytic or endophytic (Section 6.3). In general, herbivores that are more difficult for primary parasitoids to locate and successfully attack should similarly be less accessible to true hyperparasitoids. Further, if the primary parasitoids pupate in the herbivore's feeding site (e.g. within a gall or plant stem), pseudohyperparasitoid species richness should also be affected. But an additional factor important to hyperparasitoids is the prevalence of suitable koinobiont hosts in the primary parasitoid complex, which is also associated with herbivore feeding niche (Section 4.2).

This simple model, then, can account for a wide range of interactions, largely being driven by forces arising from the first and second trophic levels percolating up through the third and fourth trophic levels of plant—herbivore—parasitoid—hyperparasitoid foodwebs.

I have so far structured the model and discussion using host feeding niche as its foundation, because of the statistical associations of the various dependent variables with feeding niche. However, to extend this model beyond the statistical into the conceptual, it is clearly critical to identify the biological mechanism underlying the correlations. At several points in this monograph I have argued that feeding niche actually represents a proxy variable for the extent that hosts occupy refuges from parasitism, and the susceptibility hypothesis (Sections 3.6 and 5.6) uses these refuges as its basis. On the other hand, alternative explanations for at least some of the patterns have been proposed, so before discussing further the evidence in support of the susceptibility hypothesis, these mechanisms should be evaluated. There are currently two such alternative hypotheses.

First, there is Godfray's (1993) homogeneity hypothesis to account for the niche differences in parasitoid species richness. I have already examined this hypothesis in so far as possible with the currently available data (Sections 3.2 and 3.6) and found no support for either its taxonomic or ecological components. This would seem to eliminate this hypothesis as the primary mechanism underlying the species richness patterns. However, there is a further potential problem with Godfray's hypothesis. Even if the data did support homogeneity as the best explanation for parasitoid species richness, this hypothesis would remain low in relative 'richness' (*sensu* Slobodkin, 1986). This is because we are still left with the patterns of host mortality and biological control successes that require explanation. Taxonomic and/or ecological homogeneity seems to be unable to account for these additional results. Godfray's hypothesis could be extended to the mortality pattern if the relationship between diversity and mortality were causative; greater numbers of parasitoid species arising from homogeneity directly produce more total host mortality. Of course, this is inconsistent with the relationship between parasitoid species richness and maximum parasitism rates found in the data (Section 5.6), so this possibility is reduced. Even more important, it is difficult to see how homogeneity could produce the feeding niche–biological control relationship. Parasitoid introductions for biological control typically involve fairly specialized species directed against single host species, and the impact of the parasitoid on host population densities occurs very much in ecological time, so there is little logical basis for invoking the evolutionarily based homogeneity hypothesis. We are faced with a mechanism that is presumed to underlie one pattern but leaves other patterns unexplained. If we are going to accept this hypothesis then additional mechanisms must be proposed to account for the other patterns. If Occam's Razor applies, hypotheses that encompass all of the known patterns (i.e. are 'rich', Slobodkin, 1986) have logical precedence over any that do not. Consequently, the lack of empirical support coupled with an inability to explain the full suite of host–parasitoid patterns makes the homogeneity hypothesis an unlikely candidate as a general explanation.

A second alternative hypothesis that is much more comprehensive is Price's (1991, 1994) succession-based hypothesis, which does attempt to link parasitoid species richness, the importance of parasitoids to host population dynamics, and biological control through increasing host susceptibility as succession proceeds. Unfortunately, it is very difficult to evaluate this hypothesis for several reasons. First, some of its components are empirically founded on my own previously published analyses. Therefore, I cannot 'test' a hypothesis using data that were partially used to generate it. Second, the sheer complexity of the hypothesis makes it impossible for me to address it with the very limited number of variables for which I have data. So what, if anything, can be said about the possibility that ecological succession may account for the patterns I have found? Throughout my analyses I have examined plant effects whenever possible and have compared natural versus cultivated habitats in an effort to identify patterns in the context of succession (Sections 3.3, 3.5, 4.3, 5.3, 6.4). In at least some cases I have found evidence consistent with the idea that succession plays a role in host–parasitoid interactions. But I have also found that relationships of dependent variables with plant and habitat variables are generally much weaker and idiosyncratic than those with feeding niche, and that the latter relationships persist even when the former are incorporated into analyzes. If succession was of overriding importance, independent niche relationships would be expected to disappear once plant or habitat variables were entered into the analyses. That differences between the feeding niches are not strongly or consistently influenced suggests that, although there is support for at least some of the predictions made by Price's hypothesis, succession cannot, by itself, explain the strongest patterns. Based on the data and results I am able to bring to bear here, I conclude that mechanisms arising from ecological succession are operating on host–parasitoid systems (as are taxonomic and ecological homogeneity and a large number of other factors), but they are *relatively* weak, and an alternative mechanism must be invoked to explain the generally much stronger relationships based on herbivore feeding niches.

Even if my evaluations of the above hypotheses are substantially correct, it does not follow that the susceptibility hypothesis must be the right one. First, despite the internal consistency among the results of the large number of analyzes done, it remains possible that the relationships I have found are a fluke or arise from unknown or unrecognized directional biases in the data sets. Second, even if we accept that the patterns are a reasonable reflection of biological reality, the explanation may lie in a mechanism not yet conceived. I have attempted to check for biases in the data whenever possible, and most patterns are robust. I am unable to do more with these data. With respect to the second possibility, what evidence exists to indicate that the refuge mechanism is in fact correct?

There are several lines of support for the hypothesis, some indirect and some much more direct. First, the hypothesis is not novel. The influence of various types of refuge on host–parasitoid interactions has been widely considered in the theoretical and empirical literature, and many workers have stressed their importance (e.g. Simmonds, 1949; Murdoch & Oaten, 1975; Beddington *et al.*, 1978; Hassell, 1978; Shaw & Askew, 1979; Price *et al.*, 1980; Bauer, 1985; Murdoch, Chesson & Chesson, 1985; Price & Clancy, 1986; Murdoch *et al.*, 1987; Price, 1988; Price and Pschorn-Walcher, 1988; Walde *et al.*, 1989; Romstöck-Völkl, 1990; Gross, 1991, 1993). Based simply on the amount of evidence accumulated over the years, a refuge based hypothesis will come as no surprise to many parasitoid workers.

One approach for exploring the possible mechanisms underlaying the patterns has addressed the problem indirectly, examining the theoretical ability of refuges to produce patterns that qualitatively and quantitatively mirror the empirical, feeding niche patterns. Hochberg and Hawkins (1992, 1993), using a multispecies parasitoid model based on population-dynamic constructs, found that a proportional refuge template was sufficient to produce dome-shaped parasitoid species richness patterns and that, depending on the parameter values, reasonably good quantitative agreement between the richness of predicted and real parasitoid complexes could be obtained. On the other hand, we also found that if the feeding niche relationships are interpreted solely on the basis of structural refuges arising from plant protection, the reductions in parasitoid species predicted by the model on hosts with minimal refuges could not explain the lower species richness found on most exophytic hosts. This discrepancy arises because parasitoid species richness can be low either because hosts occupy a very small refuge or because they have a very large refuge. The only way to distinguish the possibilities is to examine parasitism rates, which in the former case will be high but will be low in the latter case. Maximum parasitism rates for holometabolous exophytics actually tend to be relatively low, indicating that their reduced parasitoid richness is not because they have no refuges but because many do possess biological, ecological and behavioral defenses against parasitoids, a point raised in various sections of this monograph. But it should be remembered that my data encompass holometabolous hosts only; that is to say, the Homoptera have not been included. At least some exophytic Homoptera (e.g. scales) do support relatively few parasitoids, suffer extremely high parasitism rates, and are frequently successfully controlled by the introduction of natural enemies during classical biological control attempts. Therefore, although even scales must have refuges (Walde *et al.*, 1989), I suspect that part of the reason that they support relatively few parasitoids (Hawkins & Lawton, 1987) reflects that, unlike many holometabolous hosts, they sit far to the left of the low–high refuge continuum.

Hochberg & Hawkins (1994) further explored the ability of refuges to account for real patterns of parasitoid species richness and found that an even simpler refuge model also produced good fits with observed parasitoid species richness. But, in this study we recognized that refuges may arise from many sources, not just from plant protection, and instead of feeding niche we used maximum percentage parasitism as an estimate of a host's maximum susceptibility to parasitoid attack, assuming that a host that may be subjected at least occasionally to 100% parasitism is highly susceptible, whereas a host never parasitized at a level greater than, say, 5% must occupy an extensive refuge or otherwise be virtually immune from attack.

The sum of the theoretical treatments to date indicates that a theory of host–parasitoid interactions based on host refuges from parasitism is plausible, and that relatively simple mathematical models can generate a wide range of patterns that are observed in data. On the other hand, the outcomes of models depend critically on the assumptions that go into them, so the refuge model cannot directly test the importance of refuges in real communities. This brings me to the final line of evidence supporting the susceptibility hypothesis.

The true strength of any hypothesis lies in the predictions that it makes. In the absence of experimental verification, the sole method for evaluating a hypothesis is to test its predictions using data that were not used to generate the hypothesis in the first place. Two such tests of predictions generated by the susceptibility hypothesis have been conducted to date.

The notion that host susceptibility might explain empirical parasitoid species richness patterns in Great Britain was first discussed by Hawkins & Lawton (1987). Hawkins & Gross (1992) subsequently reported the relationship between species richness and biological control success rates and further developed the hypothesis that refuges could be responsible. We also predicted that parasitism rates should be correlated with species richness and biological control. Importantly, this prediction was made before the global maximum parasitism rate data were available; these data were actually generated because of the prediction. As we have seen (Chapter 5), Hawkins and Gross' (1992) prediction holds.

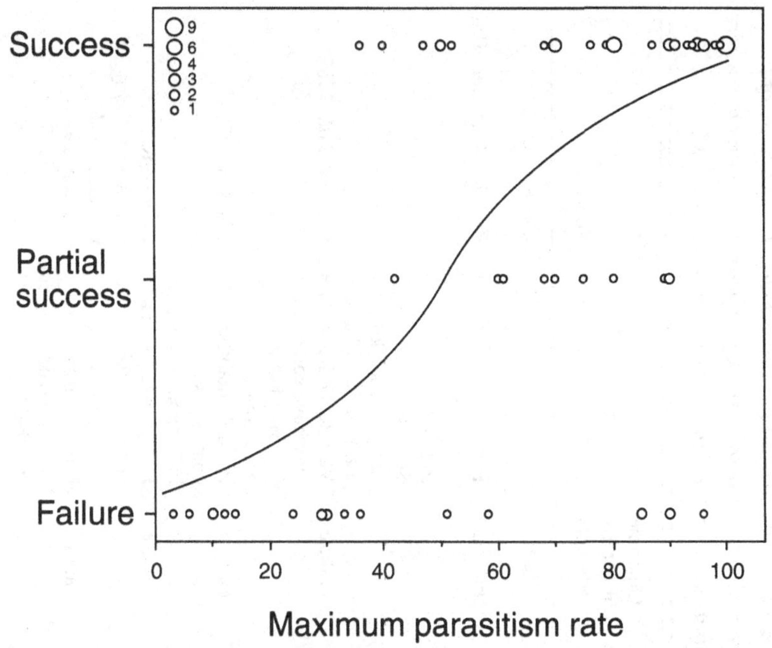

Fig. 7.2. Relationship between maximum percentage parasitism (including in two cases mortality from host feeding by parasitoids) and the outcome of parasitoid introductions for classical biological control. Multiple cases are denoted by larger dots. The regression line illustrated is based on a logistic regression when partial successes have been excluded (logity = $-2.880 + 0.057x$, $\chi^2 = 28.48$, $n = 64$, $P < 0.0000001$). Regressions in which partial success was classified as failure and in which partial success was considered success produced similar statistics (logity = $-2.957 + 0.048x$, $\chi^2 = 23.76$, $n = 74$, $P = 0.0000011$, and logity = $-2.669 + 0.058x$, $\chi^2 = 29.67$, $P < 0.0000001$, respectively). Therefore, partial successes have little influence on the overall relationship between mortality and success rates.

Hawkins, Thomas & Hochberg (1993) have recently tested a second prediction arising from refuge theory. The empirical patterns suggest an association between maximum parasitism rates and biological control success rates (Section 5.5), as does the refuge model (Hochberg and Hawkins, 1993, 1994). If the susceptibility hypothesis is correct, the extent of a host's refuge should provide a reasonable estimator of the probability that a particular parasitoid or group of parasitoids will depress host populations. Hawkins *et al.* tested this prediction using a sample of parasitoid introductions for classical biological control taken from the literature (data provided in Table 7.1) and found that refuge strength, as estimated by maximum parasitism rate, is strongly associated with the outcome of specific biological control attempts (Fig. 7.2).

Table 7.1. *List of parasitoid releases for classical biological control. The list is ranked by the maximum parasitism rate achieved by the parasitoid in any host population following parasitoid release and establishment. The outcome of the introduction is based on information provided in the sources (see notes). Target species may appear more than once where control attempts have been made at different locations or with more than one parasitoid species*

Target species	Parasitoid species	Maximum percentage parasitism	Outcome of control attempts	Notes	Source
Leucoptera coffeella (Guer.-Menv.)	*Mirax insularis* (Mues.)	3	Failure		Clausen, 1978
Hypera postica (Gyll.)	*Bathyplectes curculionis* (Thomson)+	6	Failure	Parasitoids contribute little to population trends	Kelleher & Hulme, 1984
	Tetrastichus incertus (Ratz.)			Success achieved by fungus which killed > 80% of pest larvae	
Aonidiella aurantii (Mask.)	*Prospaltella perniciosi* (Tower)	10	Failure	Very local distribution, may have some impact in spring but not at other times	Furness *et al.*, 1983
Mayetiola destructor (Say)	*Pediobius metallicus* (Nees)	10	Failure	Possible inability to compete with indigenous parasitoids	Clausen, 1978
Forficula auricularia L.	*Bigonicheta setipennis* (Fall.)	12	Failure	Pest populations are generally high	Commonwealth Institute of Biological Control, 1971
Rhabdoscelus obscurus (Boisd.)	*Lixophaga sphenophori* (Ville.)	14	Failure	Little control – compare with different sites below	Clausen, 1978

Brontispa longissima (Gastro)	*Tetrastichus brontispae* (Ferr.)	24	Failure	In New Caledonia – compare with other sites below	Cochereau, 1969
Diatraea saccharalis (F.)	*Paratheresia claripalpus* (Wulp)	29	Failure	Results disappointing – in USA in 1932 – compare with other agents and later releases below	Clausen, 1978
Neodiprion sertifer (Geoff.)	14 native and introduced species	29	Failure	In Ontario – the parasitoid introduction program contributed nothing to the control of the sawfly	Griffiths, 1959; Clausen, 1978
Oryctes rhinocerus (L.)	*Scolia ruficornis* (F.)	30	Failure	Little appreciable influence on pest populations	Clausen, 1978
Cydnia pomonella (L.)	*Ascogaster quadridentata* (Wesm.)	30	Failure	Parasitoid has not reduced key pest status	Cameron *et al.*, 1989
Eoreuma loftini (Dyar)	*Allorhogas pyralophagus* Marsh	33	Failure	Parasitism evaluated in field cages – field maximum 5%	Hawkins, Browning & Smith, 1987
Brontispa longissima	*Tetrastichus brontispae*	36	Failure	In Australia – compare with different site below	Chiu *et al.*, 1985
Nipaecoccus viridis (Newst.)	*Anagyrus indicus* Shafee *et al.* + *A. kamali* Moursi	36	Success	Infestations greatly reduced	Meyerdirk, Knasimuddin & Bashir, 1988
Operophtera brumata (L.)	*Agrypon flaveolatum* (Grav.)	40	Success	Effective at low densities – combined action with second parasitoid – see below	Clausen, 1978

Table 7.1 (*cont.*)

Target species	Parasitoid species	Maximum percentage parasitism	Outcome of control attempts	Notes	Source
Acyrthosiphon pisum (Harris)	*Aphidius eadyi* (Hall)	42	Partial success	Economic threshold in sites with parasitoids reached less frequently than in control sites	Cameron, Walker & Allan, 1981
Cephus pygmaeus (L.)	*Collyria calcitrator* (Grav.)	47	Success	In Canada – project highly successful	Clausen, 1978
Aleurocanthus woglumа (Ashby)	*Prospaltella opulenta* (Silv.)	50	Success	Good control in Texas – see other sites below	Summy & Gilstrap, 1992
Plutella xylostella (L.)	*Apanteles plutellae* (Kurdj.)	50	Success	Effective in Trinidad – see other sites below	Waterhouse & Norris, 1987
Phthorimaea operculella Zell.	*Apanteles subandinus* Blanch.	51	Failure	Parasitism insufficient to prevent large and damaging pest populations	Foot, 1979
Tryporyza nivella (F.)	*Isotima javensis* (Rohw.)	52	Success	Incidence of pest declined from 37% to 11%	Rao *et al.*, 1971
Triathaba complexa (Butler)	3 parasitoid species	58	Failure	No appreciable reduction in host population	Clausen, 1978
Coleophora laricella (Hübner)	*Chrysocharis laricinellae* (Ratz.)	60	Partial success	Control is reduced during cool periods	Commonwealth Institute of Biological Control, 1971
Popillia japonica (Newman)	*Tiphia vernalis* (Rohw.)	61	Partial success	Substantial decline in pest populations in large, open areas where adult food sources available	Clausen, 1978

Pristiphora erichsonii (Htg.)	*Mesoleius tenthredinis* Morley	68	Success	Host populations became so low that it was impractical to collect cocoons to measure parasitism rates	McLeod, 1954
Eriosoma lanigerum (Hausm.)	*Aphelinus mali* (Hald.)	68	Partial success	No full control because of slow parasitoid build up in spring	Greathead, 1976
Lamprosema octasema (Meyr.)	*Chelonus stratigenas* (Cam.)	70	Partial success		Waterhouse & Norris, 1987
Sirex noctilio F.	3 parasitoid species	70	Success	The introduction of parasitoids into New Zealand has been a success	Cameron *et al.*, 1989
Phenacoccus manihoti (Mat.-Ferr.)	*Epdinocarsis lopezi* (De Santis)	70	Success	Pest populations declined and remained low	Neuenschwander & Herren, 1988
Lepidosaphes beckii (Newman)	*Aphytis lepidosaphes* (Compere)	70	Success	In Peru – scale subsequently declined to status of secondary pest	Clausen, 1978
Proceras sacchariphagus Bojer	*Apanteles flavipes* (Cam.)	70	Success	Effect of *Apanteles* steady and satisfactory	Betededer-Matibet & Malinge, 1967
Leptinotarsus decemlineata (Say)	*Doryphorophaga doryphorae* (Riley)	75	Partial success	Effectiveness limited by poor synchronization with host	Tamaki, Chauvin & Burditt, 1983
Laspeyresia nigricana (Stephens)	*Ascogaster quadridentata*	76	Success	Reduction in infestation from 80% to 35%	Clausen, 1978
Brontispa longissima	*Tetrastichus brontispae*	79	Success	In Taiwan – see above	Chiu *et al.*, 1985
Phytomyza ilicis Curt.	4 parasitoid species	80	Partial success	Control adequate in parks, gardens and roadsides, but low economic threshold on cut holly for florists required continued insecticidal control in commercial plantings	Clausen, 1978

Table 7.1 (*cont.*)

Target species	Parasitoid species	Maximum percentage parasitism	Outcome of control attempts	Notes	Source
Operophtera brumata	*Cyzenis albicans* (Fall.)	80	Success	Effective at high densities – combined action with *A. flaveolatum* (see above)	Clausen, 1978
Phenacoccus aceris (Sign.)	*Allotropa utilis* (Mues.)	80	Success	Pest became very scarce	Clausen, 1978
Rhyacionia frustrana (Busck)	*Campoplex frustranae* (Cushm.)	80	Success	Infestations greatly reduced	Clausen, 1978
Unaspis yanonensis (Kuwana)	*Aphytis yanonensis* (DeBach & Rosen)	80	Success	Marked reduction in scale populations	Furuhashi & Okhubo, 1990
Anastrepha spp.	*Aceratoneuromyia indica* (Silv.)	80	Success	Fruit infestation reduced from 100% to 30%	Jimenez Jimenez, 1963
Phyllonorycter messaniella (Zell.)	*Apanteles circumscriptus* (Nees) + *Achrysocharoides splendens* (Delucchi)	80	Success	Infestations greatly reduced	Cameron *et al.*, 1989
Agonoxena argaula (Meyr.)	*Brachymeria agonoxenae* (Full.)	85	Failure	Heavy hyperparasitism	Clausen, 1978
Anthonomus grandis (Boh.)	*Bracon kirkpatricki* (Wlkn.)	85	Failure	Parasitoid unable to overwinter	Clausen, 1978
Antonina graminis (Mask.)	*Neodusmetia sangwani* (Rao)	87	Success	Increased yields at release sites	Gerson, Mescheloff & Dubitzki, 1975
Brontispa mariana (Spaeth)	*Tetrastichus brontispae*	89	Partial success	Full control not achieved – possible equilibrium at fairly high level	Clausen, 1978
Carulaspis minima (Targ.)	*Encarsia lounsburyi* (Berl. & Paoli)	90	Failure	Accidental introduction of agent into Bermuda	Cock, 1985

Caliroa cerasi (L.)	*Lathrolestes luteolator* (Grav.)	90	Failure	Parasitoid only attacks the less damaging of the host generations	Cameron *et al.*, 1989
Parabemisia myricae (Kuwana)	*Eretmocerus debachi* (Rose & Rosen)	90	Success	Due in part to 38% mortality from host feeding	Rose & DeBach, 1992
Leucoma salicis (L.)	*Apanteles solitarius* (Ratz.)	90	Partial success	Biological control program has not been completely effective, but considerable success is claimed	Commonwealth Institute of Biological Control, 1971
Parlatoria oleae (Colv.)	*Aphytis maculicornis* (Masi.)	90	Partial success	Pest controlled to high degree, but low economic threshold and reduced parasitoid survivorship in harsh summers meant that additional agents were required – a second parasitoid with 60% maximum parasitism improved control	DeBach, Rosen & Kennett, 1971
Rhabdoscelus obscurus	*Lixophaga spenophori*	90	Success	Borer no longer a pest – but cultural practices may be implicated in pest reduction	Rao *et al.*, 1971
Rastrococcus invadens (Williams)	*Gyranosoidea tebygi* (Noyes)	90	Success	Spectacularly successful at release point and rapidly spreading	D. Moore, personal communication
Aleurocanthus wogluma	*Eretmocerus serius* (Silv.)	91	Success	In South Africa – highly successful	Clausen, 1978
Coleophora laricella (Hübner)	*Agathis pumila* (Ratz.)	91	Success	Infestations became and remained very light	Clausen, 1978
Aleurocanthus wogluma	*Prospaltella opulenta*	93	Success	In Barbados – *Eretmocerus serius* contributed a	Clausen, 1978

Table 7.1 (*cont.*)

Target species	Parasitoid species	Maximum percentage parasitism	Outcome of control attempts	Notes	Source
				further 5% to parasitism	
Pristiphora geniculata (Htg.)	*Olesicampie geniculatae* Qued. & Lim	94	Success	Parasitoid rated as a successful agent	Kelleher & Hulme, 1984
Oulema melanopus (L.)	*Tetrastichus julis* (Walk.)	95	Success	In most fields difficult to find pest larvae	Kelleher & Hulme, 1984
Diatraea saccharalis	*Lixophaga diatraea* (T.T.)	95	Success	Joint control with *A. flavipes* below	Alam, Bennett & Carl, 1971
Diatraea saccharalis	*Apanteles flavipes*	95	Success	Joint control with *L. diatraea* above	Alam *et al.*, 1971
Mythimna separata (Walk.)	*Apanteles ruricrus* (Hal.)	95	Success	Pakistani strain brought pest under complete control	Hill, 1988
Dasyneura mali Kief.	*Prosactogaster demades* Walk.	96	Failure	Poor synchronization – maximum parasitism of second host generation 7%	Todd, 1959
Plutella xylostella	*Diadegma eucerephaga* (Grav.)	96	Success	In Indonesia – moth populations diminished appreciably	Waterhouse & Norris, 1987
Agromyza frontella (Rond.)	*Dacnusa dryas* (Nixon)	96	Success	Parasitoid caused a 50-fold reduction in host numbers	Harcourt, Guppy & Meloche, 1988
Siphoninus phillyreae Hal.	*Encarsia partenopea* (Walk.)	98	Success	Establishment of parasitoid has resulted in permanent and effective control	Gould, Bellows & Paine, 1992

Homona cofffearia (Niet.)	*Macrocentrus homonae* (Nixon)	99	Success	Pest reduced to minor status	Rao *et al.*, 1971
Chromaphis juglandicola (Kalten.)	*Trioxys pallidus* (Hal.)	100	Success	Parasitoid considered dominant factor in aphid population decline	Clausen, 1978
Lepidosaphes ficus (Sign.)	*Aphytis mytilaspidis* (Le Baron)	100	Success	Marked reduction in scale populations	Clausen, 1978
Levuana irridescens (Beth.-Baker)	*Bessa remota* (Aldrich)	100	Success	'Spectacular' success	Clausen, 1978
Nezara viridula (L.)	*Trissolcus basalis* (Woll.)	100	Success	Damage reduced to sub-economic levels	Waterhouse & Norris, 1987
Plutella xylostella	*Tetrastichus sokolowskii* (Kurd.)	100	Success	Limited effect within first season as parasitized larvae continued to feed – pest populations reduced in later crops	Waterhouse & Norris, 1987
Promecotheca coeruleipennis (Blanch.)	*Pediobius parvulus* (Ferr.)	100	Success	'Spectacular' success	Clausen, 1978
Pseudococcus citriculus (Green)	*Clausenia pupurea* (Ishii)	100	Success	Pest rare after parasitoid establishment	Clausen, 1978
Maconellicoccus hirsutus (Green)	*Anagryus kamali*	100	Success	In Egypt – in many places mealybug disappeared almost completely	Clausen, 1978
Ceroplastes rubens Maskell	*Anicetus annulatus* Ishii & Yasum.	100	Success	In Japan – agent accidentally introduced from China and intentionally spread – all releases successful	Hirose, Nakamura & Takagi, 1990

Maximum parasitism rates of less than 25% rarely, if ever, result in any control of the host, whereas rates greater than 40% have a very high probability of having at least some impact on host densities. Of course, agreement of data and theory are not perfect, because some introductions fail to achieve complete control even when maximum parasitism rates are high (i.e. host refuges are small). These 'exceptions' reflect that refuges are not the only factor influencing host–parasitoid interactions (see Table 7.1 for some of the reasons why introductions partially or completely fail to adequately control pests). Nevertheless, a basic tenet of the hypothesis that host susceptibility, measured by maximum parasitism rates, by itself provides a reasonable predictor of when parasitoids will be able to depress host densities is supported.

In sum, all of the empirical data and theory that I am able to marshal support the susceptibility hypothesis, and the idea that host refuges from parasitism form the mechanistic basis for the relationships depicted in Fig. 7.1 seems reasonable. At the risk of being repetitious, I reiterate that given the number of parasitoid and herbivorous insect species and the general complexity of ecological systems, no single factor can be expected to explain everything. I make no claim that refuges represent the only force acting on parasitoids or their hosts, or that my simple treatment of the patterns exhausts the possibilities. Further work and improved data will ultimately determine the actual contribution of refuges to the full range of host–parasitoid interactions or if alternative explanations of the patterns prove to be more robust. But based on the weight of current evidence, refuge theory does deserve much closer inspection.

It should be obvious from the structure of a refuge-based model that it represents a bottom-up interpretation of host–parasitoid interactions, in which biological, ecological and evolutionary properties of the herbivores and their foodplants dictate the size and species composition of parasitoid communities and the dynamics of the interactions among hosts and parasitoids. The role of bottom-up forces in determining community structure is an idea whose time has clearly come (Hawkins, 1992; Hunter & Price, 1992; Hunter, Ohgushi & Price, 1992; Menge, 1992; Power, 1992; Strong, 1992). This view is not new or particularly complicated; the simple observation that predators can only eat prey within a limited range of body sizes (Elton, 1927) illustrates the universally recognized constraints natural enemies are under.

It should also be obvious that bottom-up effects can arise in both evolutionary and ecological time. Of more considerable debate is the extent that natural enemies exert top-down control on victim populations in ecological time (Price, 1987; Crawley, 1992b). Refuges represent one possible mechanism to reconcile observations that some systems are under bottom-up (= donor) control whereas others are under top-down control. Refuge theory argues that a biological-control–donor-control gradient exists: when hosts occupy small refuges, parasitoids have strong effects on host populations, whereas when hosts occupy large refuges, parasitoid impact is trivial. The data do not permit a quantitative evaluation of the frequencies of top-down and bottom-up controlled dynamics in native host–parasitoid systems; but analyses of the biological control record suggest that bottom-up dynamics are more frequent than top-down dynamics when hosts and parasitoids are introduced, since fewer than half of all parasitoid introductions result in control, even considering only those cases when the parasitoids have successfully established (Hawkins, 1992). Of course, this conclusion is based on parasitoids in particular and does not address the relative importance of predators to host population dynamics. But refuge theory is also relevant to predator–prey interactions in general (e.g. Crawley, 1992a; Everett & Ruiz, 1993; Bechara, Moreau & Hare, 1993; and references therein), so this interpretation might be expected to apply similarly to both parasitoids and predators.

Finally, under refuge theory few systems should be entirely top-down or bottom-up controlled. It is only at extremely high levels of refuge that top-down effects will be entirely absent, although they may be dynamically trivial at somewhat lower refuge levels. Cases at the other extreme probably do not exist, at least for long. Hosts that have no refuge at all will quickly be driven to extinction by their parasitoids. Therefore, some bottom-up effects must occur in all host–parasitoid systems, with the possible exception of cases where hyperparasitoids depress primary parasitoid populations so severely that host populations are able to persist, even in the absence of refuges. But I doubt that such refugeless hosts are common, because herbivores will still need to escape from predators.

So, where now? My contribution to the issues involved in host–parasitoid interactions has been to take a broad-brush approach in an effort to synthesize the large amount of empirical data that has been collected over the past 100 years or so. I have tried to show two things: first, that patterns do exist and, second, that relatively simple ideas can go a long way in explaining the main features of these patterns. The 'take-home' message of this monograph is that common processes link various aspects of host–parasitoid interactions, and it is possible to fit both community- and population-wide patterns, and both basic and applied population biology, into a single conceptual framework. My hypothesis to do this is based on refuges. Time and further study will tell if this is the correct interpretation. But even if it is not, it does not alter the fact that patterns of parasitoid diversity, parasitoid-induced host mortality, and success in biological control covary in some surprising ways. The goal now is to further explore why.

It is inevitable that it has not been possible to 'explain' all of the patterns I have found or even to distinguish among several possible explanations; the comparative approach, like all methodologies used in ecology, has its limitations. Typically, the data are imperfect and fragmentary, and critical tests will require alternative approaches. Major unresolved problems revolve around species–area relationships, latitudinal gradients in parasitoid diversity, parasitoid host ranges, and the relative contributions of ecological and evolutionary forces to the interactions. I have not dealt with many important questions related to host–parasitoid interactions, such as the role of competition in parasitoid community structure (Askew & Shaw, 1986; Hawkins, 1990), parasitoid-mediated competition among herbivores (Lawton, 1986; Holt & Lawton, 1994), vacant niches (Mills, 1988; Hawkins & Compton, 1992), spatial and/or temporal heterogeneity in the intensity of interactions (Askew & Shaw, 1986; Pacala, Hassell & May, 1990), interactions among parasitoids and predators (Tostowaryk, 1971; Washburn, 1984; Tscharntke, 1992b), etc. Nor have I concentrated on details, at my own peril. There are no doubt exceptions to every conclusion I have reached, and for some patterns there will likely prove to be more exceptional cases than confirmatory ones. The detailed analyzes of the parasitoid complexes of individual host families conducted by Price & Pschorn-Walcher (1988), Mills (1994) and Hoffmeister & Vidal (1994) amply demonstrate some of the complexities involved. But no one expects it to be easy. Given the enormous diversity of the organisms that fall within the scope of the phrase 'host–parasitoid interactions', nature is bound to be complicated. On the other hand, I believe that it would be a mistake to conclude that it is too complicated to unravel. I remain an optimist, but then I am not in full possession of the facts.

References

Agricola, U. & H.-U. Fischer (1991). Hyperparasitism in two newly introduced parasitoids, *Epidinocarsis lopezi* and *Gyranuscoidea tebygi* (Hymenoptera: Encyrtidae) after their establishment in Togo. *Bulletin of Entomological Research*, **81**, 127–32.

Ahmad, R. (1974). Investigations on the white-fringed weevils *Naupactus durius* (Boh.) and *Pantomorus auripes* Hustache (Col.: Curculionidae) and their natural enemies in Argentina. *Commonwealth Institute of Biological Control Technical Bulletin*, **17**, 37–51.

Alam, M. M., F. D. Bennett & K. P. Carl (1971). Biological control of *Diatraea saccharalis* (F.) in Barbados by *Apanteles flavipes* Cam. and *Lixophaga diatraeae* T.T. *Entomophaga*, **16**, 151–8.

Andow, D. A. (1991). Vegetational diversity and arthropod population response. *Annual Review of Entomology*, **36**, 561–86.

Arnaud, P. H., Jr (1978). A host-parasite catalog of North American Tachinidae (Diptera). *United States Department of Agriculture Miscellaneous Publication*, No. 1319, 860 pp.

Askew, R. R. (1961). On the biology of the inhabitants of oak galls of Cynipidae (Hymenoptera) in Britain. *Transactions of the Society for British Entomology*, **14**, 237–68.

Askew, R. R. (1975). The organisation of chalcid-dominated parasitoid communities centred upon endophytic hosts. In *Evolutionary Strategies of Parasitic Insects and Mites*, ed. P. W. Price, pp. 130–53. New York: Plenum Press.

Askew, R. R. (1980). The diversity of insect communities in leaf-mines and plant galls. *Journal of Animal Ecology*, **49**, 817–29.

Askew, R. R. (1990). Species diversities of hymenopteran taxa in Sulawesi. In *Insects and the Rain Forests of South East Asia (Wallacea)*, ed. W. J. Knight & J. D. Holloway, pp. 255–60. London: Royal Entomological Society of London.

Askew, R. R. (1994). Parasitoids of leaf-mining Lepidoptera: what determines their host range? In *Parasitoid Community Ecology*, ed. B. A. Hawkins & W. Sheehan. Oxford: Oxford University Press.

Askew, R. R. & J. M. Ruse (1974). The biology of some Cecidomyiidae (Diptera) galling the leaves of birch (*Betula*) with special reference to their chalcidoid (Hymenoptera) parasites. *Transactions of the Royal Entomological Society of London*, **126**, 129–67.

Askew, R. R. & M. R. Shaw (1974). An account of the Chalcidoidea (Hymenoptera) parasitising leaf-mining insects of deciduous trees in Britain. *Biological Journal of the Linnean Society*, **67**, 51–64.

Askew, R. R. & M. R. Shaw (1986). Parasitoid communities: their size, structure and development. In *Insect Parasitoids*, ed. J. Waage & D. Greathead, pp. 225–64. London: Academic Press.

Ball, J. C. & D. L. Dahlston (1973). Hymenopterous parasites of *Ips paraconfusus* (Coleoptera: Scolytidae) larvae and their contribution to mortality. I. Influence of host tree and tree diameter on parasitization. *Canadian Entomologist*, **105**, 1453–64.

Barbosa, P. (1988). Natural enemies and herbivore–plant interactions: influence of plant allelochemicals and host specificity. In *Novel Aspects of Insect–Plant Interactions*, ed. P. Barbosa & D. K. Letourneau, pp. 201–30. New York: John Wiley .

Bauer, G. (1985). Strategies in parasitoid associations utilizing the same developmental stages of two rose tortricids. *Entomologia Experimentalis et Applicata*, **37**, 275–81.

Bechara, J. A., G. Moreau & L. Hare (1993). The impact of brook trout (*Salvelinus fontinalis*) on an experimental stream benthic community: the role of spatial and size refugia. *Journal of Animal Ecology*, **62**, 451–64.

Beddington, J. R., C. A. Free & J. H. Lawton (1978). Characteristics of successful natural enemies in models of biological control of insect pests. *Nature*, **273**, 513–19.

Beddington, J. R. & P. S. Hammond (1977). On the dynamics of host–parasite–hyperparasite interactions. *Journal of Animal Ecology*, **46**, 811–21.

Belshaw, R. (1994). Life history characteristics of Tachinidae (Diptera) and their effect of polyphagy. In *Parasitoid Community Ecology*, ed. B. A. Hawkins & W. Sheehan. Oxford: Oxford University Press.

Betededer-Matibet, M. & P. Malinge (1967). Un succès de la lutte biologique: controle de *Proceras sacchariphagus* Boj., 'borer ponctué' de la canne a sucre à Madagascar par un parasite introdut, *Apanteles flavipes* Cam. *Agronomie Tropicale*, **22**, 1196–220.

Blackburn, T. M. (1991). A comparative examination of life-span and fecundity in parasitoid Hymenoptera. *Journal of Animal Ecology*, **60**, 151–64.

Blackburn, T. M., V. K. Brown, B. M. Doube, J. J. D. Greenwood, J. H. Lawton & N. E. Stork (1993). The relationship between abundance and body size in natural animal assemblages. *Journal of Animal Ecology*, **62**, 519–28.

Blum, M. S. (1992). Ingested allelochemicals in insect wonderland: a menu of remarkable function. *American Entomologist*, **38**, 222–34.

Bourchier, R. S. & V. G. Nealis (1992). Patterns of hyperparasitism of *Cotesia melanoscela* (Hymenoptera: Braconidae) in southern Ontario. *Environmental Entomology*, **21**, 907–12.

Cameron, P. J., R. L. Hill, J. Bain & W. P. Thomas (1989). *A Review of Biological Control of Invertebrate Pests and Weeds in New Zealand 1874 to 1987*. Commonwealth Institute of Biological Control Technical Communication 10. Wallingford: CAB International Press.

Cameron, P. J., G. P. Walker & D. J. Allan (1981). Establishment and dispersal of the introduced parasite *Aphidius eadyi* (Hymenoptera: Aphidiidae) in the North Island of New Zealand, and its initial effect on pea aphid. *New Zealand Journal of Zoology*, **8**, 105–22.

Chiu, S. C., P. Y. Lai, B. H. Chen, Z. C. Chen & J. F. Shiau (1985). Introduction, liberation and propagation of a pupal parasitoid, *Tetrastichus brontispae*, for the control of the coconut leaf beetle in Taiwan. *Journal of Agricultural Research of China*, **34**, 213–22.

Clausen, C. P. (1940). *Entomophagous Insects*. New York: McGraw-Hill.
Clausen, C. P. (ed.) (1978). *Introduced Parasites and Predators of Arthropod Pests and Weeds: A World Review*. Agriculture Handbook No. 480. Washington, DC: United States Department of Agriculture.
Clausen, C. P., D. W. Clancy & Q. C. Chock (1965). Biological control of the oriental fruit fly (*Dacus dorsalis* Hendel) and other fruit flies in Hawaii. *United States Department of Agriculture Technical Bulletin* , No. 1322, 102 pp.
Cochereau, P. (1969). Establishment of *Tetrastichus brontispae* Ferr. (Hymenoptera, Eulophidae), parasite of *Brontispa longissima* Gastro, var. *frogatti* Sharp (Coleoptera, Chrysomelidae, Hispinae) in the Noumean peninsula. *Cahiers ORSTOM serie Biologie*, **7**, 139–41.
Cock, M. J. W. (ed.) (1985). *A Review of Biological Control of Pests in the Commonwealth Caribbean and Bermuda up to 1982*. Commonwealth Institute of Biological Control Technical Communication 9. Wallingford: CAB International Press.
Coley, P. D. & T. M. Aide (1991). Comparison of herbivory and plant defenses in temperate and tropical broad-leaved forests. In *Plant–Insect Interactions. Evolutionary Ecology in Tropical and Temperate Regions*, ed. P. W. Price, T. M. Lewinsohn, G. W. Fernandes & W. W. Benson, pp. 25–49. New York: John Wiley.
Commonwealth Institute of Biological Control (1971). *Biological Control Programmes Against Insects and Weeds in Canada 1958–1968*. Commonwealth Institute of Biological Control Technical Communication 4. Farnhan Royal: Commonwealth Agricultural Bureaux.
Cotgreave, P. (1993). The relationship between body size and population abundance in animals. *Trends in Ecology and Evolution*, **8**, 244–8.
Craig, T. P. (1994). Effects of intraspecific plant variation on parasitoid communities. In *Parasitoid Community Ecology*, ed. B. A. Hawkins & W. Sheehan. Oxford: Oxford University Press.
Crawley, M. C. (1992a). Population dynamics of natural enemies and their prey. In *Natural Enemies*, ed. M. C. Crawley, pp. 40–89. London: Blackwell Scientific.
Crawley, M. C. (1992b). Overview. In *Natural Enemies*, ed. M. C. Crawley, pp. 476–89. London: Blackwell Scientific.
DeBach, P. (1943). The importance of host-feeding by adult parasites in the reduction of host populations. *Journal of Economic Entomology*, **36**, 647–58.
DeBach, P., D. Rosen & C. E. Kennett (1971). Biological control of coccids by introduced natural enemies. In *Biological Control*, ed. C. B. Huffaker, pp. 165–94. New York: Plenum Press.
Delucchi, V. (1982). Parasitoids and hyperparasitoids of *Zeiraphera diniana* [Lep., Tortricidae] and their role in population control in outbreak areas. *Entomophaga*, **27**, 77–92.
DeVries, P. J. (1984). Butterflies and Tachinidae: does the parasite always kill its host? *Journal of Natural History*, **18**, 323–6.
Dohanian, S. M. (1942). Parasites of the filbert worm. *Journal of Economic Entomology*, **35**, 836–41.
Doutt, R. L. (1964). Biological characteristics of entomophagous adults. In *Biological Control of Insect Pests and Weeds*, ed. P. DeBach, pp. 145–67. London: Chapman & Hall.
Doutt, R. L. & P. DeBach (1964). Some biological control concepts and questions. In *Biological Control of Insect Pests and Weeds*, ed. P. DeBach, pp. 118–42. London: Chapman & Hall.
du Merle, P. (1975). Les hôtes et les stades pré-imaginaux des Diptères Bombyliidae:

revue bibliographique annotée. *Bulletin of the Western Palearctic Regional Section, International Organization for Biological Control, No.* 4, 289 pp.
Edelsten, H. M. (1933). A tachinid emerging from an adult moth. *Proceedings of the Royal Entomological Society of London*, **8**, 131.
Eggleton, P. & K. J. Gaston (1990). 'Parasitoid' species and assemblages: convenient definitions or misleading compromises? *Oikos*, **59**, 417–21.
Eggleton, P. & K. J. Gaston (1992). Tachinid host ranges: a reappraisal (Diptera: Tachinidae). *Entomologist's Gazette*, **43**, 139–43.
Ehler, L. E. (1979). Assessing competitive interactions in parasite guilds prior to introduction. *Environmental Entomology*, **8**, 558–60.
Ehler, L. E. (1990). Introduction strategies in biological control of insects. In *Critical Issues in Biological Control*, ed. M. Mackauer, L. E. Ehler & J. Roland, pp. 111–34. Andover: Intercept.
Elliott, N. C., G. A. Simmons & D. L. Haynes (1986). Mortality of pupae of jack pine budworm (Lepidoptera: Tortricidae) parasites and density dependence of hyperparasitism. *Environmental Entomology*, **15**, 662–8.
Elton, C. S. (1927). *Animal Ecology*. London: Sidgwick & Jackson.
Embree, D. G. (1974). The biological control of the winter moth in eastern Canada by introduced parasites. In *Biological Control*, ed. C. B. Huffaker, pp. 217–26. New York: Plenum Press.
English-Loeb, G. M., R. Karban & A. K. Brody (1990). Arctiid larvae survive attack by a tachinid parasitoid and produce viable offspring. *Ecological Entomology*, **15**, 361–2.
Erwin, T. L. (1982). Tropical forests: their richness in Coleoptera and other arthropod species. *Coleopterist's Bulletin*, **35**, 53–68.
Evans, E. W. (1994). Indirect interactions among phytophagous insects: aphids, honeydew, and natural enemies. In *Individuals, Populations, and Patterns in Ecology*, ed. S. R. Leather, A. D. Watt, N. J. Mills & K. F. A. Walters. Andover: Intercept.
Everett, R. A. & G. M. Ruiz (1993). Coarse woody debris as a refuge from predation in aquatic communities. *Oecologia*, **93**, 475–86.
Fahringer, J. (1941). Zur Kenntnis der Parasiten der Monne (*Lymantria monacha* L.). *Zeitschrift für Angewandte Entomologie*, **28**, 335–58.
Feeny, P. (1976). Plant apparency and chemical defence. *Recent Advances in Phytochemistry*, **10**, 1–40.
Flanders, S. E. (1963). Hyperparasitism, a mutualistic phenomenon. *Canadian Entomologist*, **95**, 716–20.
Foot, M. A. (1979). Bionomics of the potato tuber moth, *Phthorimaea operculella* (Lepidoptera: Gelechiidae), at Pukekohe. *New Zealand Journal of Zoology*, **6**, 623–36.
Force, D. C. (1970). Competition among four hymenopterous parasites of an endemic host. *Annals of the Entomological Society of America*, **63**, 1675–88.
Force, D. C. (1980). Do parasitoids of Lepidoptera larvae compete for hosts? Probably. *American Naturalist*, **116**, 873–5.
Force, D. C. (1985). Competition among parasitoids of endophytic hosts. *American Naturalist*, **126**, 440–4.
Furness, G. O., G. A. Buchanan, R. S. George & N. L. Richardson (1983). A history of the biological and integrated control of red scale, *Aonidiella aurantii* on citrus in the lower Murray Valley of Australia. *Entomophaga*, **28**, 199–212.
Furuhashi, K. & N. Okhubo (1990). Use of parasitic wasps to control arrowhead scale, *Unaspis yanonensis* (Homoptera, Diaspididae) in Japan. In *The Use of Natural Enemies to Control Agricultural Pests*, ed. O. Mochida, K. Kiritani & J. Bay-

Petersen, pp. 71–82. Tapei: Food and Fertilizer Technology Center for the Asian and Pacific Region Book Series, No. 40.

Garthwaite, P. F. & M. H. Desai (1939). On the biology of the parasites of the teak defoliator, *Hapalia machaeralis* Walk. (Pyralidae) and *Hyblaea puera* Cram. (Hyblaeidae) in Burma. *Indian Forest Records*, **5**, 309–53.

Gaston, K. J. & J. H. Lawton (1988). Patterns in body size, population dynamics, and regional distribution of bracken herbivores. *American Naturalist*, **132**, 662–80.

Gaston, K. J., D. Reavey & R. G. Valladares (1991). Changes in feeding habit as caterpillars grow. *Ecological Entomology*, **16**, 339–44.

Gauld, I. D. (1986). Latitudinal gradients in ichneumonid species-richness in Australia. *Ecological Entomology*, **11**, 155–61.

Gauld, I. D. (1987). Some factors affecting the composition of tropical ichneumonoid faunas. *Biological Journal of the Linnean Society*, **30**, 299–312.

Gauld, I. D. (1988). Evolutionary patterns of host utilization by ichneumonoid parasitoids (Hymenoptera: Ichneumonidae and Braconidae). *Biological Journal of the Linnean Society*, **35**, 351–77.

Gauld, I. & B. Bolton (1988). *The Hymenoptera*. Oxford: Oxford University Press.

Gauld, I. D. & K. J. Gaston (1994). The taste of enemy-free space: parasitoids and nasty hosts. In *Parasitoid Community Ecology*, ed. B. A. Hawkins & W. Sheehan. Oxford: Oxford University Press.

Gauld, I. D., K. J. Gaston & D. H. Janzen (1992). Plant allelochemicals, tritrophic interactions and the anomalous diversity of tropical parasitoids: the 'nasty' host hypothesis. *Oikos*, **65**, 353–7.

Gerson, U., E. Mescheloff & E. Dubitzki (1975). The introduction of *Neodusmetia sangwani* (Rao) (Hymenoptera: Encyrtidae) into Israel for the biological control of the rhodesgrass scale, *Antonina graminis* (Maskell) (Homoptera: Pseudococcidae). *Journal of Applied Ecology*, **12**, 767–99.

Godfray, H. C. J. (1993). *Parasitoids: Behavioral and Evolutionary Ecology*. Princeton: Princeton University Press.

Gordh, G. (1981). The phenomenon of insect hyperparasitism and its taxonomic occurrence in the Insecta. In *The Role of Hyperparasitism in Biological Control: A Symposium*, ed. D. Rosen, pp. 10–18. Priced Publication 4103. Berkeley: Division of Agricultural Sciences, University of California.

Gothilf, S. (1969). Natural enemies of the carob moth *Ectomyelois cereatoniae* (Zeller). *Entomophaga*, **14**, 195–202.

Gould, J. R., T. S. Bellows, Jr & T. D. Paine (1992). Population dynamics of *Siphoninus phllyreae* in California in the presence and absence of a parasitoid, *Encarsia parenopea*. *Ecological Entomology*, **17**, 127–43.

Greathead, D. (ed.) (1976). *A Review of Biological Control in Western and Southern Europe*. Commonwealth Institute of Biological Control Technical Communication 7. Farnhan Royal: Commonwealth Agricultural Bureaux.

Greathead, D. (1986). Parasitoids in classical biological control. In *Insect Parasitoids*, ed. J. Waage & D. Greathead, pp. 289–318. London: Academic Press.

Greathead, D. & A. H. Greathead (1992). Biological control of insect pests by insect parasitoids and predators: the BIOCAT database. *Biocontrol News and Information*, **13**, 61N–8N.

Griffiths, K. J. (1959). Observations on the European pine sawfly, *Neodiprion sertifer* (Geoff.), and its parasites in southern Ontario. *Canadian Entomologist*, **91**, 501–12.

Gross, P. (1991). Influence of target feeding niche on success rates in classical biological control. *Environmental Entomology*, **20**, 1217–27.

Gross, P. (1993). Insect behavioral and morphological defenses against parasitoids. *Annual Review of Entomology*, **38**, 251–73.

Gutierrez, A. P. & R. van den Bosch (1970). Studies on host selection and host specificity of the aphid hyperparasite *Charips victrix* (Hymenoptera: Cynipidae). 2. The bionomics of *Charips victrix*. *Annals of the Entomological Society of America*, **63**, 1355–60.

Haeselbarth, E. (1979). Zur Parasitierung der Puppen von Forleuhe (*Panolis flammea* (Schiff.)), Kiefernspanner (*Bupalus piniarius* (L.)) und Heidelbeerspanner (*Boarmia bistortana* (Goezel)) in bayerischen Kiefernwäldern. *Zeitschrift für Angewandte Entomologie*, **87**, 186–202, 311–22.

Hagen, K. S. (1964). Developmental stages of parasites. In *Biological Control of Insect Pests and Weeds*, ed. P. DeBach, pp. 168–246. London: Chapman & Hall.

Hairston, N. G., Sr (1989). *Ecological Experiments. Purpose, Design, and Execution*. Cambridge: Cambridge University Press.

Hall, R. W., L. E. Ehler & B. Bisabri-Ershadi (1980). Rates of success in classical biological control of arthropods. *Bulletin of the Entomological Society of America*, **26**, 111–14.

Harcourt, D. G., J. C. Guppy & F. Meloche (1988). Population dynamics of the alfalfa blotch leafminer, *Agromyza frontella* (Diptera: Agromyzidae), in eastern Ontario: impact of the exotic parasite *Dacnusa dryas* (Hymenoptera: Braconidae). *Environmental Entomology*, **17**, 337–43.

Harman, D. M. & H. M. Kulman (eds.) (1973). *A World Survey of the Parasites and Predators of the Genus Rhyacionia, Parts I to IV*, Contribution No. 527, pp. 79–178. University of Maryland Natural Research Institute.

Harvey, P. H. & M. D. Pagel (1991). *The Comparative Method in Evolutionary Biology*. Oxford: Oxford University Press.

Hassell, M. P. (1978). *Arthropod Predator–Prey Systems*. Princeton: Princeton University Press.

Hassell, M. P. (1986). Parasitoids and population regulation. In *Insect Parasitoids*, ed. J. Waage & D. Greathead, pp. 201–24. London: Academic Press.

Hassell, M. P. & J. K. Waage (1984). Host–parasitoid population interactions. *Annual Review of Entomology*, **29**, 89–114.

Hawkins, B. A. (1988). Species diversity in the third and fourth trophic levels: patterns and mechanisms. *Journal of Animal Ecology*, **57**, 137–62.

Hawkins, B. A. (1990). Global patterns of parasitoid assemblage size. *Journal of Animal Ecology*, **59**, 57–72.

Hawkins, B. A. (1992). Parasitoid–host food webs and donor control. *Oikos*, **65**, 159–62.

Hawkins, B. A. (1993a). Parasitoid species richness, host mortality, and biological control. *American Naturalist*, **141**, 634–41.

Hawkins, B. A. (1993b). Refuges, host population dynamics, and the genesis of parasitoid diversity. In *Hymenoptera and Biodiversity*, ed. J. LaSalle & I. D. Gauld, pp. 235–56. Wallingford: CAB International Press.

Hawkins, B. A., R. R. Askew & M. R. Shaw (1990). Influences of host feeding-niche and foodplant type on generalist and specialist parasitoids. *Ecological Entomology*, **15**, 275–80.

Hawkins, B. A., H. W. Browning & J. W. Smith, Jr (1987). Field evaluation of *Allorhogas pyralophagus* [Hym.: Braconidae], imported into Texas for biological control of the stalkborer *Eoreuma loftini* [Lep.: Pyralidae] in sugar cane. *Entomophaga*, **32**, 483–91.

Hawkins, B. A. & S. G. Compton (1992). African fig wasp communities: undersaturation and latitudinal gradients in species richness. *Journal of Animal Ecology*, **61**, 361–72.

Hawkins, B. A. & R. J. Gagné (1989). Determinants of assemblage size for the parasitoids of Cecidomyiidae (Diptera). *Oecologia*, **81**, 75–88.

Hawkins, B. A. & R. D. Goeden (1984). Organization of a parasitoid community associated with a complex of galls on *Atriplex* spp. in southern California. *Ecological Entomology*, **9**, 271–92.
Hawkins, B. A. & P. Gross (1992). Species richness and population limitation in insect parasitoid–host systems. *American Naturalist*, **139**, 417–23.
Hawkins, B. A. & J. H. Lawton (1987). Species richness for parasitoids of British phytophagous insects. *Nature*, **326**, 788–90.
Hawkins, B. A. & J. H. Lawton (1988). Species richness patterns: why do some insects have more parasitoids than others. In *Parasitoid Insects*, ed. M. Bouletreau & G. Bonnot, pp. 131–6. Les Colloques de l'INRA, No. 48. Paris: Institut National de la Recherche Agronomique.
Hawkins, B. A., M. R. Shaw & R. R. Askew (1992). Relationships among assemblage size, host specialization, and climatic variability in North American parasitoid communities. *American Naturalist*, **139**, 58–79.
Hawkins, B. A., M. B. Thomas & M. E. Hochberg (1993). Refuge theory and biological control. *Science*, **262**, 1429–37.
Hespenheide, H. A. (1979). Are there fewer parasitoids in the tropics? *American Naturalist*, **113**, 766–9.
Hill, M. G. (1988). Analysis of the biological control of *Mythimna separata* (Lepidoptera: Noctuidae) by *Apanteles ruficrus* (Hymenoptera: Braconidae) in New Zealand. *Journal of Applied Ecology*, **25**, 197–208.
Hirose, Y. (1994). Determinants of species richness and composition in egg parasitoid assemblages of Lepidoptera. In *Parasitoid Community Ecology*, ed. B. A. Hawkins & W. Sheehan. Oxford: Oxford University Press.
Hirose, Y., T. Nakamura & M. Takagi (1990). Successful biological control: a case study of parasite aggregation. In *Critical Issues in Biological Control*, ed. M. Mackauer, L. E. Ehler & J. Roland, pp. 171–83. Andover: Intercept.
Hochberg, M. E. & B. A. Hawkins (1992). Refuges as a predictor of parasitoid diversity. *Science*, **225**, 973–6.
Hochberg, M. E. & B. A. Hawkins (1993). Predicting parasitoid species richness. *American Naturalist*, **142**, 671–93.
Hochberg, M. E. & B. A. Hawkins (1994). The implications of population dynamics theory to parasitoid diversity and biological control. In *Parasitoid Community Ecology*, ed. B. A. Hawkins & W. Sheehan. Oxford: Oxford University Press.
Hochberg, M. E. & J. H. Lawton (1990). Competition between kingdoms. *Trends in Ecology and Evolution*, **5**, 367–71.
Hoffmeister, T. & S. Vidal (1994). The diversity of fruit fly (Diptera: Tephritidae) parasitoids. In *Parasitoid Community Ecology*, ed. B. A. Hawkins & W. Sheehan. Oxford: Oxford University Press.
Hokkanen, H. M. T. (1986). Success in classical biological control. *CRC Critical Reviews in Plant Science*, **3**, 35–72.
Holt, R. D. & J. H. Lawton (1994). Apparent competition and enemy-free space in insect host–parasitoid communities. *American Naturalist*, in press.
Huffaker, C. B., P. S. Messenger & P. DeBach (1974). The natural enemy component in natural control and the theory of biological control. In *Biological Control*, ed. C. B. Huffaker, pp. 16–67. London: Plenum Press.
Hunter, M. D., T. Ohgushi & P. W. Price (ed) (1992). *Effects of Resource Distribution on Animal–Plant Interactions*. San Diego: Academic Press.
Hunter, M. D. & P. W. Price (1992). Playing chutes and ladders: heterogeneity and the relative roles of bottom-up and top-down forces in natural communities. *Ecology*, **73**, 724–32.

Isart, J. (1972). Observaciones bioecológicas sobre *Cleonus mendicus* (Gyllenhal, 1834) (Col. Curculionidae). *Graellsia*, **28**, 177–201.
Janzen, D. H. (1975). Interactions of seeds and their insect predators/parasitoids in a tropical deciduous forest. In *Evolutionary Strategies of Parasitic Insects and Mites*, ed. P. W. Price, pp. 154–86. New York: Plenum Press.
Janzen, D. H. (1981). The peak in North American ichneumonid species richness lies between 38° and 42° N. *Ecology*, **62**, 532–7.
Janzen, D. H. & C. M. Pond (1975). A comparison, by sweep sampling, of the arthropod fauna of secondary vegetation in Michigan, England and Costa Rica. *Transactions of the Royal Entomological Society of London*, **127**, 33–50.
Jervis, M. A. & N. A. C. Kidd (1986). Host-feeding strategies in hymenopteran parasitoids. *Biological Reviews*, **61**, 395–434.
Jimenez Jimenez, E. (1963). Avances y resultados del control biologico en Mexico. *Fitofilo*, No. 38, pp. 34–7.
Jones, D. A., J. Parsons & M. Rothschild (1962). Release of hydrocyanic acid from crushed tissues of all stages in the life-cycle of species of the Zygaeninae (Lepidoptera). *Nature*, **193**, 52–53.
Keddy, P. A. (1989). *Competition*. London: Chapman & Hall.
Kelleher, J. S. & M. A. Hulme (1984). *Biological Control Programmes Against Insects and Weeds in Canada, 1909–1980*. Farnhan Royal: Commonwealth Agricultural Bureaux.
Keller, M. A. (1984). Reassessing evidence for competitive exclusion of introduced natural enemies. *Environmental Entomology*, **13**, 192–5.
Kloet, G. S. & W. D. Hincks (1972). *A Checklist of British Insects. Part 2. Lepidoptera*. London: Royal Entomological Society of London.
Kloet, G. S. & W. D. Hincks (1976). *A Checklist of British Insects. Part 5. Diptera and Siphonaptera*. London: Royal Entomological Society of London.
Kloet, G. S. & W. D. Hincks (1977). *A Checklist of British Insects. Part 3. Coleoptera and Strepsiptera*. London: Royal Entomological Society of London,
Kloet, G. S. & W. D. Hincks (1978). *A Checklist of British Insects. Part 4. Hymenoptera*. London: Royal Entomological Society of London.
Krishna Ayyar, P. N. (1940). The alternate host plants and associated parasites of *Pempheres affinis* Faust in south India. *Indian Journal of Entomology*, **2**, 213–27.
Lampe, K.-H. (1984). Struktur und Dynamik des Parasitenkomplexes der Binsensackträgermotte *Coleophora alticolella* Zeller (Lep.: Coleophoridae) in Mitteleuropa. *Zoologische Jahrbüch, Abteilung für Systematik, Ökologie und Geographie der Tiere*, **111**, 449–92.
LaSalle, J. & I. D. Gauld (1992). Parasitic Hymenoptera and the biodiversity crisis. *Redia*, **74** (Appendice), 315–34.
Lawton, J. H. (1983). Plant architecture and the diversity of phytophagous insects. *Annual Review of Entomology*, **28**, 23–9.
Lawton, J. H. (1986). The effect of parasitoids on phytophagous insect communities. In *Insect Parasitoids*, ed. J. Waage & D. Greathead, pp. 265–87. London: Academic Press.
Lawton, J. H. (1989). Food webs. In *Ecological Concepts. The Contribution of Ecology to an Understanding of the Natural World*, ed. J. M. Cherrett, pp. 43–78. Oxford: Blackwell Scientific.
Lawton, J. H. (1990). Species richness and population dynamics of animal assemblages. Patterns in body size:abundance space. *Philosophical Transactions of the Royal Society of London B*, **330**, 283–91.
Le Pelley, R. H. (1937). Notes on the life-history of *Cheiloneurus noxius*, Compere (Hym.). *Bulletin of Entomological Research*, **28**, 181–3.

Luck, R. M., P. S. Messenger & J. F. Barbieri (1981). The influence of hyperparasitism on the performance of biological control agents. In *The Role of Hyperparasitism in Biological Control: A Symposium*, ed. D. Rosen, pp. 34–42. Priced Publication 4103. Berkeley: Division of Agricultural Sciences, University of California.

Matthews, R. W. (1974). Biology of Braconidae. *Annual Review of Entomology*, **19**, 15–32.

May, R. M. & M. P. Hassell (1981). The dynamics of multiparasitoid–host interactions. *American Naturalist*, **117**, 234–61.

Mayer, R. P. & J. K. Shull (1978). *Tachinobia repanda*: a hyperparasite of the *Megalopyge opercularis–Carcelia lagoae* association. *Florida Entomologist*, **61**, 241–3.

McDaniel, J. R. & V. C. Moran (1972). The parasitoid complex of the citrus psylla *Trioza erytreae* (Del Guercio) (Homoptera: Psyllidae). *Entomophaga*, **17**, 297–317.

McLeod, J. H. (1954). Statuses of some introduced parasites and their hosts in British Columbia. *Proceedings of the Entomological Society of British Columbia*, **50** (1953), 19–27.

Memmott, J. & H. J. C. Godfray (1994). The use and construction of parasitoid webs. In *Parasitoid Community Ecology*, ed. B. A. Hawkins & W. Sheehan. Oxford: Oxford University Press.

Menge, B. A. (1992). Community regulation: under what conditions are bottom-up factors important on rocky shores? *Ecology*, **73**, 755–65.

Meyerdirk, D. E., S. Khasimuddin & M. Bashir (1988). Importation, colonization and establishment of *Anagyrus indicus* [Hym. Encyrtidae] on *Nipaecoccus viridis* [Hom.: Pseudococcidae] in Jordan. *Entomophaga*, **33**, 229–37.

Miller, J. C. (1980). Niche relationships among parasitic insects occurring in a temporary habitat. *Ecology*, **61**, 270–5.

Miller, J. C. & L. E. Ehler (1990). The concept of parasitoid guilds and its relevance to biological control. In *Critical Issues in Biological Control*, ed. M. Mackauer, L. E. Ehler & J. Roland, pp. 159–69. Andover: Intercept.

Miller, P. F. (1973). The biology of some *Phyllonoryctor* species (Lepidoptera: Gracillariidae) mining leaves on oak and beech. *Journal of Natural History*, **7**, 391–409.

Mills, N. J. (1988). The structure and diversity of the parasitoid complexes of *Zeiraphera* bud moths in relation to the dynamics of their populations. In *Parasitoid Insects*, ed. M. Bouletreau & G. Bonnot, pp. 159–60. Les Colloques de l'INRA, No. 48. Paris: Institut National de la Recherche Agronomique.

Mills, N. J. (1992). Parasitoid guilds, life-styles, and host ranges in the parasitoid complexes of tortricoid hosts (Lepidoptera: Tortricidae). *Environmental Entomology*, **21**, 320–9.

Mills, N. J. (1993). Species richness and structure in the parasitoid complexes of tortricoid hosts. *Journal of Animal Ecology*, **62**, 45–58.

Mills, N. J. (1994). Parasitoid guilds: a comparative analysis of the parasitoid communities of tortricids and weevils. In *Parasitoid Community Ecology*, ed. B. A. Hawkins & W. Sheehan. Oxford: Oxford University Press.

Mills, N. J. & M. Kenis (1991). A study of the parasitoid complex of the European fir budworm, *Choristoneura muriana* (Lepidoptera: Tortricidae), and its relevance for biological control of related hosts. *Bulletin of Entomological Research*, **81**, 429–36.

Milne, A. (1963). Biology and ecology of the garden chafer, *Phyllopertha horticola* (L.). IX. Spatial distribution. *Bulletin of Entomological Research*, **54**, 761–95.

Morris, R. F. (1976). Hyperparasitism in populations of *Hyphantria cunea*. *Canadian Entomologist*, **108**, 685–7.
Morrison, G., M. Auerbach & E. D. McCoy (1979). Anomalous diversity of tropical parasitoids: a general phenomenon? *American Naturalist*, **114**, 303–7.
Muesebeck, C. F. W. & S. M. Dohanian (1927). A study in hyperparasitism, with particular reference to the hyperparasites of *Apanteles melanoscelus* (Ratzeburg). *United States Department of Agriculture Department Bulletin* No. 1487, 35 pp.
Murdoch, W. W., J. Chesson & P. L. Chesson (1985). Biological control in theory and practice. *American Naturalist*, **125**, 344–66.
Murdoch, W. W., R. M. Nisbet, S. P. Blythe, W. S. C. Curney & J. D. Reeve (1987). An invulnerable age class and stability in delay-differential parasitoid–host models. *American Naturalist*, **129**, 263–82.
Murdoch, W. W. & A. Oaten (1975). Predation and population stability. *Advances in Ecological Research*, **9**,1–131.
Mustata, G. (1978). Facteurs biotiques limitatifs dans certaines populations de *Chortophila brassicae* Bouché, (Diptera, Muscidae). *Travaux du Museum d'Histoire Naturelle 'Grigore Antipa'*, **19**, 289–91.
Myers, J. H., C. Higgens & E. Kovacs (1989). How many insect species are necessary for the biological control of insects. *Environmental Entomology*, **18**, 541–7.
Nagarkatti, S. & K. Ramachandran Nair (1973). The influence of wild and cultivated Gramineae and Cyperaceae on populations of sugarcane borers and their parasites in north India. *Entomophaga*, **18**, 419–30.
Narayanan, E. S. &. B. R. Subba Rao (1960). Super-, multi- and hyperparasitism and their effect on the biological control of insect pests. *Proceedings of the National Institute of Science of India B*, **26** (suppl.), 257–80.
Neuenschwander, P. & P. Herren (1988). Biological control of the cassava mealybug, *Phenacoccus manihoti*, by the exotic parasitoid *Epidinocarsis lopez* in Africa. *Philosophical Transactions of the Royal Philosophical Society of London B*, **318**, 319–33.
Nicholson, A. J. & V. A. Bailey (1935). The balance of animal populations. Part I. *Proceedings of the Zoological Society of London*, 1935, 551–98.
Noyes, J. S. (1989). The diversity of Hymenoptera in the tropics with special reference to Parasitica in Sulawesi. *Ecological Entomology*, **14**, 303–7.
Owen, D. F. & J. Owen (1974). Species diversity in temperate and tropical Ichneumonidae. *Nature*, **249**, 583–4.
Pacala, S. W., M. P. Hassell & R. M. May (1990). Host–parasitoid associations in patchy environments. *Nature*, **344**, 150–3.
Paine, R. T. (1983). Intertidal food webs: does connectance describe their essence? In *Current Trends in Food Web Theory. Report on a Food Web Workshop*, ed. D. L. DeAngelis, W. M. Post & G. Sugihara, pp. 11–16. Oak Ridge, TN: Oak Ridge National Laboratory.
Paine, R. T. (1992). Food-web analysis through field measurement of per capita interaction strength. *Nature*, **355**, 73–5.
Peters, R. H. (1991). *A Critique for Ecology*. Cambridge: Cambridge University Press.
Petersen, J. J. (1986). Evaluating the impact of pteromalid parasites on filth fly populations associated with confined livestock installations. *Miscellaneous Publications of the Entomological Society of America*, **61**, 52–6.
Petersen, J. J., M. A. Catangui & D. W. Watson (1991). Parasitoid-induced mortality of house fly pupae by pteromalid wasps in the laboratory. *Biological Control*, **1**, 275–80.
Pianka, E. R. (1978). *Evolutionary Ecology*. New York: Harper & Row.
Pike, K. S. & C. C. Burkhardt (1974). Hyperparasites of *Bathyplectes curculionis* in Wyoming. *Environmental Entomology*, **3**, 953–6.

Power, M. E. (1992). Top-down and bottom-up forces in food webs: do plants have primacy? *Ecology*, **73**, 733–46.
Price, P. W. (1980). *Evolutionary Biology of Parasites*. Princeton: Princeton University Press.
Price, P. W. (1987). The role of natural enemies in insect populations. In *Insect Outbreaks*, ed. P. Barbosa & J. C. Schultz, pp. 287–312. London: Academic Press.
Price, P. W. (1988). Inversely density-dependent parasitism: the role of plant refuges for hosts. *Journal of Animal Ecology*, **57**, 89–96.
Price, P. W. (1991). Evolutionary theory of host and parasitoid interactions. *Biological Control*, **1**, 83–93.
Price, P. W. (1994). Evolution of parasitoid communities. In *Parasitoid Community Ecology*, ed. B. A. Hawkins & W. Sheehan. Oxford: Oxford University Press.
Price, P. W., C. E. Bouton, P. Gross, B. A. McPheron, J. N. Thompson & A. E. Weis (1980). Interactions among three trophic levels: influence of plants on interactions between insect herbivores and natural enemies. *Annual Review of Ecology and Systematics*, **11**, 41–65.
Price, P. W. & K. M. Clancy (1986). Interactions among three trophic levels: gall size and parasitic attack. *Ecology*, **67**, 1593–1600.
Price, P. W. & H. Pschorn-Walcher (1988). Are galling insects better protected against parasitoids than exposed feeders?: a test using tenthredinid sawflies. *Ecological Entomology*, **13**, 195–205.
Pschorn-Walcher, H. & E. Altenhofer (1989). The parasitoid community of leaf-mining sawflies (Fenusini and Heterarthrini): a comparative analysis. *Zoologischer Anzeiger*, **222**, 37–56.
Rao, V. P., M. A. Ghani, T. Sankaran & K. C. Mathur (1971). *A Review of the Biological Control of Insects and Other Pests in South-East Asia and the Pacific Region.* Commonwealth Institute of Biological Control Technical Communication 6. Farnhan Royal: Commonwealth Agricultural Bureaux.
Rasplus, J.-Y. (1994). Parasitoid communities associated with west-African seed-feeding beetles. In *Parasitoid Community Ecology*, ed. B. A. Hawkins & W. Sheehan. Oxford: Oxford University Press.
Rathcke, B. J. & P. W. Price (1976). Anomalous diversity of tropical ichneumonid parasitoids: a predation hypothesis. *American Naturalist*, **110**, 889–93.
Rathman, R. J. & T. F. Watson (1985). A survey of early-season host plants and parasites of *Heliothus* spp. in Arizona. *Journal of Agricultural Entomology*, **2**, 388–94.
Rohde, K. (1992). Latitudinal gradients in species diversity: the search for the primary cause. *Oikos*, **65**, 514–27.
Romstöck-Völkl, M. (1990). Host refuges and spatial patterns of parasitism in an endophytic host–parasitoid system. *Ecological Entomology*, **15**, 321–31.
Rose, M. & P. DeBach (1992). Biological control of *Parabemisia myricae* (Kuwana) (Homoptera: Aleyrodidae) in California. *Israel Journal of Entomology*, **25–26**, 73–96.
Rosen, D. (ed.) (1981). *The Role of Hyperparasitism in Biological Control: A Symposium*. Priced Publication 4103. Berkeley: Division of Agricultural Sciences, University of California.
Rothschild, M. (1985). British aposematic Lepidoptera. In *The Moths and Butterflies of Great Britain and Ireland. Vol. 2 Cossidae–Heliodinidae*, ed. J. Heath & A. M. Emmet, pp. 9–62. Colchester: Harley Books.
Salt, G. (1963). The defence reactions of insect metazoan parasites. *Parasitology*, **53**, 527–642.

Sandanayake, W. R. M. & J. P. Edirisinghe (1992). *Trathala falvoorbitalis*: parasitization and development in relation to host-stage attacked. *Insect Science and its Application*, **13**, 287–92.

Sato, H. (1990). Parasitoid complexes of lepidopteran leaf miners on oaks (*Quercus dentata* and *Quercus mongolica*) in Hokkaido, Japan. *Ecological Research*, **5**, 1–8.

Schaffner, J. V. (1959). Microlepidoptera and their parasites reared from field collections in the northeastern United States. *United States Department of Agriculture Miscellaneous Publication* No. 767, 97 pp.

Schaffner, J. V. & C. L. Griswold (1934). Macrolepidoptera and their parasites reared from field collections in the northeastern part of the United States. *United States Department of Agriculture Miscellaneous Publication* No. 188, 160 pp.

Schönrogge, K. & E. Altenhofer (1992). On the biology and larval parasitoids of the leaf-mining sawflies *Profenusa thomsoni* (Konow) and *P. pygmaea* (Konow) (Hym., Tenthredinidae). *Entomologist's Monthly Magazine*, **128**, 99–108.

Schröder, D. (1974). A study of the interactions between the internal larval parasites of *Rhyacionia buoliana* (Lepidoptera: Olethreutidae). *Entomophaga*, **19**, 145–71.

Shaw, M. R. (1993). Species of *Mesochorus* (Hymenoptera: Ichneumonidae) reared as hyperparasitoids of Lepidoptera via koinobiont ectoparasitoid Tryphoninae (Ichneumonidae). *Entomologist's Gazette*, **44**, 181–82.

Shaw, M. R. (1994). Parasitoid host ranges. In *Parasitoid Community Ecology*, ed. B. A. Hawkins & W. Sheehan. Oxford: Oxford University Press.

Shaw, M. R. & J.-P. Aeschlimann (1994). Host ranges of parasitoids (Hymenoptera: Braconidae and Ichneumonidae) reared from *Epermentia chaerophyllella* (Goeze) (Lepidoptera: Epermeniidae) in Britain, with description of a new species of *Triclistus* (Ichneumonidae). *Journal of Natural History*, in press.

Shaw, M. R. & R. R. Askew (1976a). Ichneumonoidea (Hymenoptera) parasitic upon leaf-mining insects of the orders Lepidoptera, Hymenoptera and Coleoptera. *Ecological Entomology*, **1**, 127–33.

Shaw, M. R. & R. R. Askew (1976b). Parasites. In *The Moths and Butterflies of Great Britain and Ireland. Vol. 1. Micropterigidae–Heliozelidae*, ed. J. Heath, pp. 24–56. Oxford: Blackwell Scientific.

Shaw, M. R. & R. R. Askew (1979). Hymenopterous parasites of Diptera (Hymenoptera Parasitica). In *A Dipterist's Handbook*, 15. ed. A. Stubbs & P. Chandler, pp. 164–71. Hanworth: The Amateur Entomologist's Society.

Shaw, M. R. & T. Huddleston (1991). Classification and Biology of Braconid Wasps (Hymenoptera: Braconidae). *Handbook for the Identification of British Insects*, Vol. 7, Part 11. London: Royal Entomological Society of London.

Sheehan, W. (1986). Response by specialist and generalist natural enemies to agroecosystem diversification: a selective review. *Environmental Entomology*, **15**, 456–61.

Sheehan, W. (1991). Host range patterns of hymenopteran parasitoids of exophytic lepidopteran folivores. In *Insect–Plant Interactions*, Vol. III, ed. E. Bernays, pp. 209–47. Boca Raton, FL: CRC Press.

Sheehan, W. & B. A. Hawkins (1991). Attack strategy as an indicator of host range in metopiine and pimpline Ichneumonidae (Hymenoptera). *Ecological Entomology*, **16**, 129–31.

Sickle, D. & R. M. Weseloh (1974). Habitat variables that influence the attack by hyperparasites of *Apanteles melanoscelus* cocoons. *Journal of the New York Entomological Society*, **82**, 2–5.

Simmonds, F. J. (1949). Some difficulties in determining by means of field samples the true value of parasitic control. *Bulletin of Entomological Research*, **39**, 435–40.

Slobodkin, L. B. (1986). The role of minimalism in art and science. *American Naturalist*, **127**, 257–65.
Smith, J. W. Jr, R. N. Wiedenmann & W. A. Overholt (1993). *Parasites of Lepidopteran Stemborers of Tropical Gramineous Plants*. Nairobi: ICIPE Press.
Southwood, T. R. E. (1976). Habitat, the template for ecological strategies? *Journal of Animal Ecology*, **46**, 337–65.
Stamp, N. E. (1981). Behavior of parasitized aposematic caterpillars: advantageous to the parasitoid or the host? *American Naturalist*, **118**, 715–25.
Stevens, G. C. (1989). The latitudinal gradient in geographical range: how so many species coexist in the tropics. *American Naturalist*, **133**, 240–56.
Stiling, P. (1990). Calculating the establishment rates of parasitoids in classical biological control. *American Entomologist*, **36**, 225–30.
Strong, D. R. (1992). Are trophic cascades all wet? Differentiation and donor-control in speciose ecosystems. *Ecology*, **73**, 747–54.
Strong, D. R., J. H. Lawton & R. Southwood (1984). *Insects on Plants. Community Patterns and Mechanisms*. Oxford: Blackwell Scientific.
Sullivan, D. J. (1987). Insect hyperparasitism. *Annual Review of Entomology*, **32**, 49–70.
Summy, K. R. & F. E. Gilstrap (1992). Regulation of citrus blackfly (Homoptera: Aleyrodidae) by *Encarsia opulenta* (Hymenoptera: Aphelinidae) on Texas citrus. *Biological Control*, **2**, 19–27.
Tamaki, G., R. L. Chauvin & A. K. Burditt, Jr (1983). Field evaluation of *Doryphorophaga doryphorae* (Diptera: Tachinidae), a parasite, and its host the Colorado potato beetle (Coleoptera: Chrysomelidae). *Environmental Entomology*, **12**, 386–9.
Todd, D. H. (1959). The apple leaf-curling midge, *Dasyneura mali* Kieffer, seasonal history, varietal susceptibility and parasitism 1955–58. *New Zealand Journal of Agricultural Research*, **2**, 859–69.
Tostowaryk, W. (1971). Relationship between parasitism and predation of diprionid sawflies. *Annals of the Entomological Society of America*, **64**, 1424–7.
Towner, C. (1992). Community ecology of phytophagous insects and their parasitoids. Ph.D. Dissertation, University of London.
Townes, H. (1971). Ichneumonidae as biological control agents. *Proceedings of the Tall Timbers Conference on Ecological Animal Control by Habitat Management*, **3**, 235–48.
Tscharntke, T. (1992a). Coexistence, tritrophic interactions and density dependence in a species-rich parasitoid community. *Journal of Animal Ecology*, **61**, 59–67.
Tscharntke, T. (1992b). Cascade effects among four trophic levels: bird predation on galls effects density-dependent parasitism. *Ecology*, **73**, 1689–98.
van Alphen, J. J. M. & L. E. M. Vet (1986). An evolutionary approach to host finding and selection. In *Insect Parasitoids*, ed. J. Waage & D. Greathead, pp. 23–61. London: Academic Press.
van den Bosch, R. (1981). Specificity of hyperparasites. In *The Role of Hyperparasitism in Biological Control: A Symposium*, ed. D. Rosen, pp. 27–33. Priced Publication 4103. Berkeley: Division of Agricultural Sciences, University of California.
Van Driesche, R. G. (1983). The meaning of 'percent parasitism' in studies of insect parasitoids. *Environmental Entomology*, **12**, 1611–22.
Van Driesche, R. G., T. S. Bellows, Jr, J. S. Elkinton, J. R. Gould & D. N. Ferro (1991). The meaning of percentage parasitism revisited: solutions to the problem of accurately estimating total losses from parasitism. *Environmental Entomology*, **20**, 1–7.

van Emden, F. I. (1954). *Diptera Cyclorrapha, Calyptrata (1) section (a): Tachinidae and Calliphoridae*. Handbooks for the Identification of British Insects 10. London: Royal Entomological Society of London.
van Emden, H. F. (1981). Wild plants in the ecology of insect pests. In *Pests, Pathogens and Vegetation*, ed. J. M. Thresh, pp. 251–61. London: Pitman.
van Emden, H. F. (1990). Plant diversity and natural enemy efficiency in agroecosystems. In *Critical Issues in Biological Control*, ed. M. Mackauer, L. E. Ehler & J. Roland, pp. 63–80. Andover: Intercept.
Vinson, S. B. (1981). Habitat location. In *Semiochemicals: Their Role in Pest Control*, ed. D. A. Nordlund, R. L. Jones & W. J. Lewis, pp. 51–77. New York: John Wiley.
Waage, J. (1990). Ecological theory and the selection of biological control agents. In *Critical Issues in Biological Control*, ed. M. Mackauer, L. E. Ehler & J. Roland, pp. 135–57. Andover: Intercept.
Walde, S. J., R. F. Luck, D. S. Yu & W. W. Murdoch (1989). A refuge for red scale: the role of size-selectivity by a parasitoid wasp. *Ecology*, **70**, 1700–6.
Washburn, J. O. (1984). Mutualism between a cynipid gall wasp and ants. *Ecology*, **65**, 654–6.
Waterhouse, D. R. & K. R. Norris (1987). *Biological Control: Pacific Prospects*. Melbourne: Inkata Press.
Way, M. J. & M. E. Cammell (1981). Effects of weeds and weed control on invertebrate pest ecology. In *Pests, Pathogens and Vegetation*, ed. J. M. Thresh, pp. 443–58. London: Pitman.
Weseloh, R. M. (1979). Competition among gypsy moth hyperparasites attacking *Apanteles melanoscelus*, and influence of temperature on their field activity. *Environmental Entomology*, **8**, 86–90.
Weseloh, R. M. (1983). Population sampling method for cocoons of the gypsy moth (Lepidoptera: Lymantriidae) parasite, *Apanteles melanoscelus* (Hymenoptera: Braconidae), and relationship of its population levels to predator- and hyperparasite-induced mortality. *Environmental Entomology*, **12**, 1228–31.
Weseloh, R. M. (1986). Hyperparasitoids of the gypsy moth (Lepidoptera: Lymantriidae): field attack patterns on *Cotesia melanoscela* (Hymenoptera: Braconidae) at different host densities and on different-sized host clumps. *Annals of the Entomological Society of America*, **79**, 308–11.
Wharton, R. A. (1993). Bionomics of the Braconidae. *Annual Review of Entomology*, **38**, 121–43.
Wilson, L. F. (1968). Life history, habits and damage of a gall midge, *Oligotrophus papyriferae* (Diptera: Cecidomyiidae), injurious to paper birch in Michigan. *Canadian Entomologist*, **100**, 777–84.
Wishart, G., E. H. Colhoun & A. E. Monteith (1957). Parasites of *Hylemya* spp. (Diptera: Anthomyiidae) that attack cruciferous crops in Europe. *Canadian Entomologist*, **89**, 510–17.
World Conservation Monitoring Centre (1992). *Global Biodiversity: Status of the Earth's Living Resources*. London: Chapman & Hall.
Zwölfer, H. (1979). Strategies and counterstrategies in insect population systems competing for space and food in flower heads and plant galls. *Fortschritte der Zoologie*, **25**, 331–53.

Index